Petroleum Chemistry and Refining

Applied Energy Technology Series

James G. Speight, Ph.D., *Editor*

Khan **Conversion and Utilization of Waste Materials**
Khartchenko **Advanced Energy Systems**
Lee **Alternative Fuels**
Mushrush and Speight **Petroleum Products: Instability and Incompatibility**
Speight **Environmental Technology Handbook**
Speight **Petroleum Chemistry and Refining**
Wyman **Handbook on Bioethanol: Production and Utilization**

Petroleum Chemistry and Refining

Edited by

James G. Speight
Western Research Institute
Laramie, Wyoming, U.S.A.

Taylor & Francis
Publishers since 1798

USA	Publishing Office:	Taylor & Francis 1101 Vermont Avenue, NW, Suite 200 Washington, DC 20005-3521 Tel: (202) 289-2174 Fax: (202) 289-3665
	Distribution Center:	Taylor & Francis 1900 Frost Road, Suite 101 Bristol, PA 19007-1598 Tel: (215) 785-5800 Fax: (215) 785-5515
UK		Taylor & Francis Ltd. 1 Gunpowder Square London EC4A 3DE Tel: 0171 583 0490 Fax: 0171 583 0581

PETROLEUM CHEMISTRY AND REFINING

1 2 3 4 5 6 7 8 9 0 BRBR 9 0 9 8 7

This book was set in Times Roman by James G. Speight. The acqusition editor was Lisa Ehmer. The editor was Alison Labbate. Cover design by Ed Berg, Atkeson Design.

A CIP catalog record for this book is available from the British Library.
∞ The paper in this publication meets the requirements of the ANSI Standard Z39.48-1984 (Permanence of Paper)

Library of Congress Cataloging-in-Publication Data

Petroleum chemistry and refining / James G. Speight
p. cm. -- (Applied energy technology series)
Includes bibliographical references and index.
1. Petroleum - Refining. 2. Petroleum chemicals. I. Title. II. Series
TP690.S744 1998
665.5--dc21 98-17306
ISBN 1-56032-587-9

This book is dedicated to the memory of Bob Long (1923–1996)

Colleague and Friend

Contents

Contributors

Geoffrey E. Dolbear
G. E. Dolbear and Associates
23050 Aspen Knoll Drive
Diamond Bar, CA 91765

M. Rashid Khan
Texaco Fuels & Lubricants Technology Department
P.O. Box 509
Beacon, NY 12508

Robert B. Long (deceased)

George W. Mushrush
Department of Chemistry
George Mason University
Fairfax, VA 22030-4444

David J. Patmore
National Center for Upgrading Technology
CANMET Western Research Laboratory
1 Oil Patch Drive
Suite A202
Devon, Alberta
Canada T9G 1A8

John G. Reynolds
University of California
Lawrence Livermore National Laboratory
P. O. Box 808, L-369
Livermore, CA 94551

James G. Speight
Western Research Institute
365 North 9th Street,
Laramie, WY 82070-3380

Preface

The use of petroleum is an old art but the petroleum industry is essentially a 20th century industry, and without a powerful boost from chemistry, the full potential of petroleum might never have been realized. The first use of petroleum came over five thousand years ago when it was recognized that the heavier derivatives of petroleum (asphalt) could be used for caulking, as adhesives for jewelry or for construction purposes, as well as for medicinal and other purposes. However, it was its use as an illuminant that brought petroleum to the fore.

The kerosene fraction became the fraction of choice and the first refineries were essentially distillation units to isolate as much of the kerosene fraction as possible. The introduction of cracking (thermal decomposition) as applied in other industries allowed the early refiners to maximize the production of kerosene. And it was through chemistry that an understanding of the production of kerosene and other products through thermal the search to increase the amounts of illuminant materials And it is through chemistry in conjunction with other disciplines that petroleum will continue to be used to its maximum potential.

Over the past two decades the *average* quality of petroleum delivered to refineries has diminished by approximately two points on the API gravity scale. This has been accompanied by, among other things, an increase in the sulfur content. As a result, refineries have had to adapt processing strategies to deliver liquid fuel products from feedstocks that are deteriorating in quality. As the reserves of lighter crude oils continue to decline, there is increasing need to develop options to upgrade the abundant supply of known heavy oil reserves into more value-added products. Thus, there is an ever-increasing need to understand the chemical nature of the constituents and the chemistry of conversion processes.

This book is an extension of a tutorial entitled *Petroleum Chemistry and Refining* sponsored by the Petroleum Division and held at the beginning of the American Chemical Society meeting in Chicago in August 1995.

It was felt by the contributors that there is a need to educate the newer inductees into refinery science and technology about the nature of petroleum and the chemistry and physics that occur during refinery operations. Over the past two decades, there has also been a decrease in the numbers of scientists and engineers who are assigned the task of investigating the chemical and physical properties of petroleum and their effects on refining. Thus, there is a knowledge gap between the established scientists and engineers and the newer scientists and engineers who are showing an interest in refinery science and technology.

To summarize the book, *Chapter 1* deals with the *composition of petroleum* and shows how formation affects composition. Methods for determining composition are given culminating in the presentation of *composition maps* that can be employed to predict properties, understand separation processes, and illustrate refinery processes. *Chapter 2* deals with *petroleum analysis and evaluation* and focuses on the various standard methods that can be used to evaluate petroleum for refining. *Chapter 3* gives details of the *metals* (more specifically nickel and vanadium) and *heteroatoms* (nitrogen, oxygen, and sulfur) in petroleum that have deleterious effects on catalysts during processing. *Chapter 4* is an outline of the *asphaltic constituents* (specifically the asphaltene fraction) of petroleum and describes how this fraction behaves during refining, and *Chapter 5* is a follow-up to Chapter 4 and deals with the *thermal chemistry of petroleum constituents*. *Chapter 6* summarizes *refinery processes*, particularly those processes that might be applied to the heavy feedstocks, and *Chapter 7* focuses on *hydrocracking chemistry* with emphasis on *catalysts and processes*. Finally, *Chapter 8* deals with the phenomenon of *incompatibility*, which is becoming more prevalent in petroleum products and during heavy crude oil refining. Each chapter is supported by numerous references that the reader can use for more details.

The intended audience for this book is wide ranging and includes the newer inductees into petroleum science and technology as well as those already involved in this important field of science and engineering who need a broader coverage of the subject area.

And finally, a tribute to a colleague and friend, Bob Long, who died while the book was in preparation. It is always difficult to remember the death of a close friend. It is always a pleasure to remember the friendship that has endured over the years. Bob undertook, and completed, the task of co-authoring a chapter with his usual gusto and relish. He brought to the book several decades of experience with the Exxon Research and Engineering Company. It is for these contributions that he will be remembered by many scientists and engineers. He will always be remembered by those of us who could call him friend and whose lives he touched. It is to Bob's memory that this book is dedicated. *Requiescat in pace*.

Dr. James G. Speight
Editor

Chapter 1

The Composition of Petroleum

Robert B. Long (deceased)

and

James G. Speight
Western Research Institute
365 North 9th Street
Laramie
Wyoming 82072

1.0 INTRODUCTION

Petroleum was first used over five thousand years ago when it was recognized that the heavier derivatives of petroleum (asphalt) could be used for caulking and as adhesives for jewelry or for construction purposes. There is also documented use of asphalt for medicinal purposes. Approximately two thousand years ago, Arabian scientists developed methods for the distillation of petroleum that were introduced into Europe by way of Spain. Interest in petroleum was also documented in China, where petroleum was encountered when drilling for salt, and appears in documents of the third century AD. The Baku region of northern Persia was also reported (by Marco Polo in 1271–1273) to have a commercial petroleum industry. Of course, it must be realized that there may be vast differences between commercial operations in the thirteenth century and commercial operations in the twentieth century!

Interest in petroleum continued up to modern times with an increasing interest in nafta (naphtha) when it was discovered that this material could be used as an illuminant and as a supplement to asphalt incendiaries in warfare. The nafta of that time was obtained from shallow wells or by the destructive distillation of asphalt. This may perhaps be equated to the modern-day coking operations (Chapter 6), in which the overall objective of the process is to convert the nonvolatile residua into liquid fuels. The modern oil industry began in 1859 with the discovery and subsequent commercialization of petroleum in Pennsylvania (Bell, 1945). It is from this time that the modern petroleum industry evolved into its present form.

Refining petroleum involves subjecting the feedstock to a series of physical and chemical processes (Figure 1-1) as a result of which a variety of products are generated. In some of the processes, e.g., distillation, the constituents of the feedstock are isolated unchanged, whereas in other processes, e.g., cracking, considerable molecular changes to the constituents are brought about.

In the early days of refining, the results of the application of such processes to the feedstocks had a higher degree of predictability. However, the changing nature of refinery feedstocks has required that the industry evolve to accommodate the new refining chemistry. In this time period, the *average* quality of petroleum delivered to refineries has diminished by approximately two points on the API gravity scale. This has been accompanied by, among other things, an increase in the sulfur content.

The use of the term *average* is open to definition, because some refineries accept medium- to high-quality crude oil while others accept low- to medium-quality crude oil. Nevertheless, the general quality of the crude oil accepted by refineries is lower than it was two decades ago. As a result, refineries have had to adapt processing strategies to deliver liquid fuel products from feedstocks that are deteriorating in quality. The same might be stated for asphalt products. There is a shortage of high-quality asphalt because of the changing slate of refinery feedstocks.

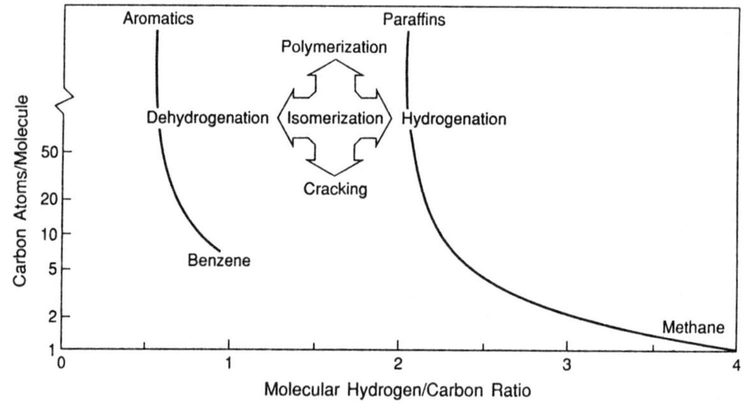

Figure 1-1. Illustration of the chemical changes that occur during petroleum refining.

Thus, as the reserves of lighter crude oils continue to decline, there is increasing need to develop options to upgrade the abundant supply of known heavy oil reserves into more value-added products. As a result, it has become more important than ever that the composition of the different feedstocks be investigated to give the refiner indications of the processing options that may be required and to afford some degree of predictability.

Recognition that refinery behavior is related to the composition of the feedstock has led to a multiplicity of attempts to establish petroleum and its fractions as compositions of matter. As a result, various analytical techniques have been developed for the identification and quantification of every molecule in the lower boiling fractions of petroleum.

Unfortunately for these past efforts, however, it is now generally recognized that the name *petroleum* does not describe a composition of matter but rather a mixture of various organic compounds that includes a wide range of molecular weights and molecular types.

Because of the complex nature of petroleum, there has been a need to define petroleum in a more meaningful and understandable manner. The general terminology that has been applied to petroleum science and technology has left much to be desired. The terminology has been inherited from the early days of petroleum use and is often misleading. Although there is still much to be accomplished, some success has been achieved. For example, it is now rare that asphaltic deposits are referred to (environmental issues notwithstanding) as *slime pits*!

2.0 DEFINITIONS

The definitions of petroleum and associated materials are varied, to the point of being archaic, and can lead to much confusion. The terms *slime* and *tar* in ancient texts often referred to the higher boiling constituents of petroleum that remained after evaporation of the lower boiling constituents at a surface seepage.

The definitions that remain in current use, for the most part, came into being during the latter part of the nineteenth century and the early part of the twentieth century at a time when the nomenclature of organic chemistry was not truly formalized. In summary, these definitions are descriptions of convenience and not necessarily accurate in a scientific sense.

Petroleum, also known as *crude oil*, is the product of natural changes to organic debris over millennia. It is a water-white to black liquid that may vary from free flowing to having difficulty in being mobile at room temperature. The predominant constituents of petroleum are hydrocarbons, but compounds containing nitrogen, oxygen, sulfur, and metals are also present. Thus, a simple definition of *petroleum* (also known as *crude oil*) is a naturally occurring mixture of hydrocarbons in a gaseous, liquid, or solid state. Such a definition might also include natural gas and, with some leeway, the mineral waxes, although the latter are usually defined separately.

Heavy oils are the less mobile (higher viscosity) constituents of the petroleum family whereas *bitumen* is even less mobile (higher viscosity) than heavy oil and is the soluble organic matter which occurs in tar sand deposits such as those in Canada, the United States, and many other places in the world (Berkowitz and Speight, 1975; Phizackerley and Scott, 1979; Kuuskraa and Hammershaimb, 1984). Bitumen is currently recovered commercially by flotation with hot water (Berkowitz and Speight, 1975). Throughout this chapter, the term *heavy oil* will be used in the generic sense to include *bitumen*.

A *residuum* or *resid* (pl. *residua*, *resids*) is the nondistillable fraction of a petroleum (Speight, 1991). Distillation temperatures are ordinarily limited to temperatures below the onset of thermal degradation (350°C; 660°F) but it is possible to recover gas oils with true boiling points up to 565°C (1050°F) by vacuum distillation. In the past, residua were used primarily as industrial fuels or asphalt precursors. The shifting product demand to lighter products coupled with the use of heavier petroleum, which has high proportions of residua, has focused attention on conversion of residua to distillate products. An *asphalt* is the product of a refinery operation whereby a residuum is treated by air blowing or by a solvent method to produce a product that meets specifications for a variety of road/highway construction and other uses.

In the simplest sense, petroleum and related materials can be considered to be composites of four major fractions that are often, for convenience, referred to using names such as *saturates*, *aromatics*, *resins* (or *polar aromatics*), and *asphaltenes* (Figure 1-2). However, chemically speaking, these names may not be truly indicative of all of the chemical constituents of the fractions, because their use lies within the historical development of petroleum science and the fraction names are operational and related more to the general characteristics than to the identification of specific compound types. Once a convenient fractionation technique has been established, it is possible to compare (Corbett and Petrossi, 1978; Speight 1991) a variety of different feedstocks varying from conventional petroleum to heavy oil as well as to make comparisons on a before-and-after-refining basis.

There are also two other operational terms that should be noted at this point: *carbenes* and *carboids* (Figure 1-2). Both such fractions are, by definition, insoluble in benzene (or toluene), but the carbenes are soluble in carbon disulfide whereas the carboids are insoluble in carbon disulfide. Only traces of these

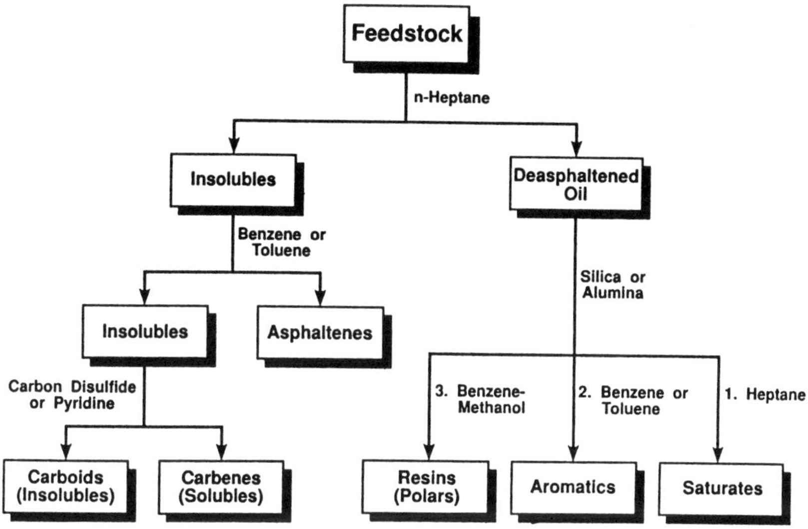

Figure 1-2. Petroleum fractionation.

materials occur in petroleum and heavy oil, and the quantities are not sufficient to alter the character of the asphaltenes if separation from the asphaltenes is omitted. On the other hand, feedstocks that have received some thermal treatment (such as visbroken feedstocks) may have considerable quantities of carbenes and carboids, as they are also considered to be precursors to coke.

Finally, the terms *deasphalting* and *deasphaltening* tend to be used interchangeably, with the former being more commonly used to mean the removal of an asphaltene fraction from petroleum by the addition of a low-boiling hydrocarbon liquid. It would be more correct to use the term *deasphalting* to indicate the removal of an asphalt-type (tacky, semisolid) material from petroleum as occurs in a refinery asphalt plant when liquid propane or liquid butane is used as the precipitant. On this basis, the term *deasphaltening* should be used to describe the removal of a solid powdery asphaltene fraction from petroleum as occurs during the addition of hydrocarbons such as pentane and heptane that are liquid under ambient conditions.

3.0 PETROLEUM FORMATION

Understanding the events by which petroleum is formed can be extremely valuable in understanding the variation in composition and properties of petroleum, which, in turn, is indicative of the behavior of petroleum in refining operations.

First, and by way of an additional definition that is specific to this section, *protopetroleum* is a generic term that has been employed on occasion to indicate the product after initial changes in the precursors have occurred that result in the formation of petroleum. In some instances, the terms *protopetroleum* and *kerogen* have been used interchangeably, although there is the notion that *protopetroleum* is the first product of diagenesis and the kerogen is a later product of this sequence.

Thus, using this form of terminology, differences in petroleum composition can be ascribed not only to the nature of the precursors that form the protopetroleum but also to the relative amounts of precursors in the mix and the maturation conditions under which the protopetroleum is converted to kerogen and thence to petroleum.

Petroleum is generally accepted as being formed from buried marine sediments by the action of heat and pressure. Therefore, its composition depends on (1) the nature of the sediment; (2) the underground environment in terms of temperature, pressure, and inorganic catalysts present with the sediment; (3) naturally occurring separation processes working on the petroleum formed; and (4) underground migration of liquids and gases to places where they are trapped into reservoirs.

Marine sediment is a term used to describe the organic biomass believed to be the raw material from which petroleum is derived, and it is a mixture of many types of marine organic material that collected at the bottom of the seas and then became buried by the geological action of the earth. The types of marine organic material that collected in the sediment could be bacteria, plankton, animals, fish, and marine vegetation in varying proportions in the different sediments buried at various locations around the world. Thus, the elemental analysis of the different masses of marine sediments buried around the world differed greatly in hydrogen-to-carbon ratio and sulfur, nitrogen, oxygen and metals content. In addition, the components of the marine sediment differed greatly in their reactivity and the types of reactions they would readily undergo.

These buried marine deposits then undergo a series of concurrent and consecutive chemical reactions collectively called *diagenesis* under the influence of the temperature, pressure, and long reaction times afforded by history in the earth. Because the sediments were buried at different locations and at different depths around the world, the resulting environment for diagenesis could vary greatly in temperature, pressure, and catalyst content. In addition to the diagenetic changes that occur in the original biomass deposit, the product can be further converted either thermally or catalytically as part of the *maturation process*.

The crust of the earth contains acidic and basic minerals as well as many metal-containing ores that can catalyze a wide variety of organic reactions capable of either changing the hydrocarbon structures and/or increasing or decreasing molecular weight and/or inserting or removing heteroatoms from the organic structure. Obviously, the amounts and types of these catalysts are not the same for all locations around the earth, and this, taken together with variations in subsurface temperature and pressure, indicates the potential for a wide variety of conversion reactions in the formation of different oil fields.

The *separation processes* that take place underground during the movement of petroleum liquids and gases through the formations are a result of the adsorptive properties of various buried minerals as well as the solvent properties of various petroleum fractions and supercritical water at the temperatures and pressures encountered underground. The general result of such effects is to change to composition of the fluids (Speight, 1991).

The pools of oil and gas that collect in geologic traps (reservoirs) are the result of fluid *migration*. These reservoirs are usually a considerable distance from the source rock, both vertically and horizontally, and it is often difficult to locate the source rocks for a given reservoir of oil or gas.

Among the products of the *conversions* and separations mentioned above, the light (less dense, lower boiling) products tend to move rapidly through the underground structure whereas the heavier (more dense, higher boiling) liquids move slowly and the solids remain in the source rock, unless carried by a solvent. Furthermore, the adsorptive power of certain minerals either immobilizes or slows the motion of adsorbed molecules. The outcome is the separation of the protopetroleum and petroleum in a manner analogous to the laboratory-type chromatographic separation by which petroleum is subdivided into a variety of fractions.

Another proposal is that the organic debris also forms a material known as *kerogen* from which petroleum is eventually formed by the various maturation processes (Tissot and Welte, 1978; Durand, 1980). However, the exact mechanism by which petroleum is formed from kerogen is unclear and speculative (Figure 1-3); there are several theories, and the proponents of each theory can vociferously state their reasons.

Thus, the nature of the petroleum that is recovered from a well depends on the ratios of the various precursors that form the protopetroleum after which the effects that are collectively called diagenisis come into operation. Whether or not the protopetroleum forms kerogen prior to the formation of petroleum is another part of the speculative puzzle. Nevertheless, physical effects that can occur during

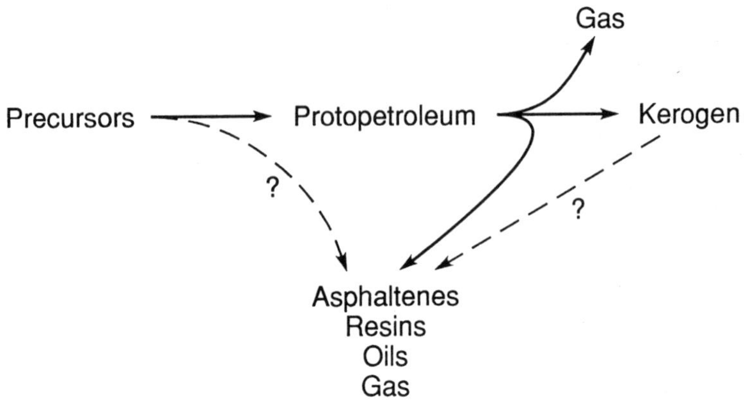

Figure 1-3. Representation of the pathways by which petroleum is formed.

migration from the source bed to the reservoir also influence the composition of petroleum.

In consideration of these effects, it is perhaps remarkable that petroleum has such a narrow spread of elemental (utlimate) compositon (Speight, 1991):

Element	Range, wt. %
carbon	83.0–87.0
hydrogen	10.0–14.0
nitrogen	0.1–2.0
oxygen	0.05–1.5
sulfur	0.05–6.0

However, it is not so much the elemental composition (which may be a reflection of the physical or fractional composition) of petroleum that determines its behavior and properties. It is the fractional composition of petroleum, more specifically the differences in petroleum composition of the crude oils, that determines the properties and behavior.

In many cases, the differences in properties can be ascribed to differences in the ratios of the various hydrocarbon constituents, recalling that not all hydrocarbons (compounds of carbon and hydrogen only) have the same properties. Cyclohexane and benzene can be cited as six-carbon compounds that have widely different physical and chemical properties. In addition, and perhaps more pertinent, the heteroatom compounds have properties that differ substantially from those of the hydrocarbons .

Thus, petroleum is formed from various precursors and (whether it be by way of a mysterious material called protopetroleum or a somewhat less mysterious material called kerogen) it is the relative amounts of these precursors as well as the prevailing maturation conditions that dictate the composition of petroleum. And it is the composition of petroleum that dictates its behavior in refineries.

4.0 PROPERTIES AND GENERAL CHARACTERISTICS

Petroleum has been generally classified as a hydrocarbon resource. On occasion, coal and oil shale kerogen have been included (somewhat erroneously) in this classification. However, the inclusion of coal and oil shale under the broad classification of hydrocarbon resources has required that the term *hydrocarbon* be expanded to include (erroneously from the chemical viewpoint) all of the non-hydrocarbon heteroatomic species in coal and oil shale kerogen. Use of the term organic sediments would be more correct (Figure 1-4). The inclusion of coal and oil shale kerogen in the catagory *hydrocarbon resources* is due to the fact that these two natural resources (coal and oil shale kerogen) will produce hydrocarbons on high-temperature processing. Therefore, if coal and oil shale kerogen are to be included in the term *hydrocarbon resources*, it is more appropriate that they be classed as *hydrocarbon-producing resources* under the general classification of *organic sediments* (Figure 1-5).

Petroleum contains high proportions of individual hydrocarbons (Bestougeff, 1967), whereas coal and oil shale kerogen contain high proportions of heteroatomic species (Scouten, 1990). The mineral waxes, such as ozocerite, can also be shown on this scheme (Figure 1-5) but, because of their character (solid), fall at the lower end of the scale. This should not be construed to mean that the mineral waxes are the same heteroatomic materials as coal and oil shale kerogen but is a result of their physical state.

Petroleum varies in appearance from a thin (mobile), nearly colorless, liquid to a thick (viscous), almost black oil. The specific gravity (at 60°F, 15.6°C) varies correspondingly from about 0.75 to 1.00 (57 to 10° API), with the specific gravity of most crude oils falling in the range 0.80 to 0.95 (45 to 17° API) Thus, it is not surprising that petroleum varies in composition from one oil field to another, from one well to another in the same field, and even from one level to another in the same well. This variation can be in both molecular weight and the types of molecules present in petroleum. Thus, petroleum may well be described as a

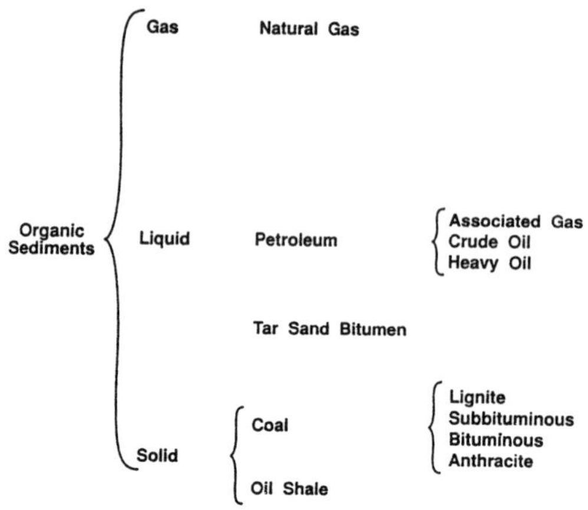

Figure 1-4. Subdivision of organic sediments.

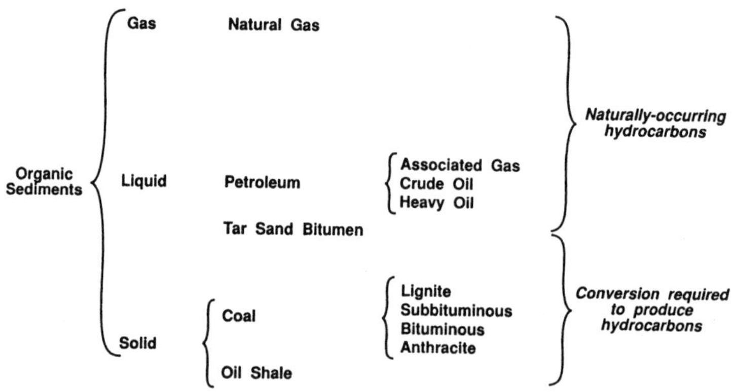

Figure 1-5. Classification of fossil fuels as naturally hydrocarbon resources and as hydrocarbon-producing resources as a result of conversion.

mixture of organic molecules drawn from a wide distribution of molecular types that lie within a wide distribution of molecular weights.

As a result of the wide variety of organic constituents and physical conditions that can play a role in the formation of petroleum, it is not surprising that there exists a wide variation in composition and properties.

On a relative basis, petroleum and heavy oil can be assessed or defined by their relative distillation characteristics (Table 1-1), but the properties and character of a residuum from one particular petroleum depend upon the temperature at which the distillation of the volatile constituents is terminated. Nevertheless, these materials are very difficult to classify since the constituents of heavy oils and residua fall into a boiling range in which very little is known about

Table 1-1. Properties of Tia Juana crude oil and its 650°F, 950°F, and 1050°F residua (Speight, 1991).

	Whole crude	Residua		
		650°F$^+$	950°F$^+$	1050°F$^+$
Yield, vol. %	100.0	48.9	23.8	17.9
Sulfur, wt. %	1.08	1.78	2.35	2.59
Nitrogen, wt. %		0.33	0.52	0.60
API gravity	31.6	17.3	9.9	7.1
Carbon residue, wt. %				
Conradson		9.3	17.2	21.6
Metals				
vanadium, ppm		185		450
nickel, ppm		25		64
Viscosity				
Kinematic				
at 100°F	10.2	890		
at 210°F		35	1010	7959
Furol				
at 122°F		172		
at 210°F			484	3760
Pour point, °F	-5	45	95	120

the individual model compounds (Figure 1-6). It is even more confusing when an increase in boiling point is directly equated to an increase in the number of rings without acknowledging that an increase in boiling point of any ring compound can also (and more likely in the case of petroleum constituents) be equated to an increase in the number or size of the substituents on the ring (Figure 1-7).

In terms of physical and chemical properties, there are several correlations that can be made between the various properties of petroleum, heavy oil, and residua that emphasize the relatively poor quality of the heavier feedstocks. For example, heavy oils and residua generally have a high proportion of sulfur and nitrogen as well as a high thermal coke-forming propensity (i.e., carbon residue) and a high viscosity. The general properties of any of these feedstocks will give an *indication* of the problems that may occur during the refining of these materials, and there is a need for properties that will indicate more *definitive* aspects of the behavior of these feedstocks. It is this latter area in which a knowledge of the composition of feedstocks can play a major role in defining the chemical nature of petroleum constituents, the means by which they interact with each other, the means by which they form coke, or the means by which they react with a catalyst and cause deactivation.

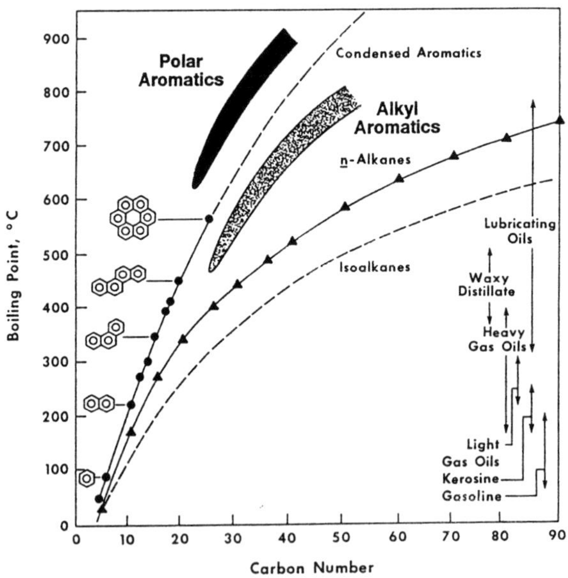

Figure 1-6. Increase in boiling point with carbon number for hydrocarbons.

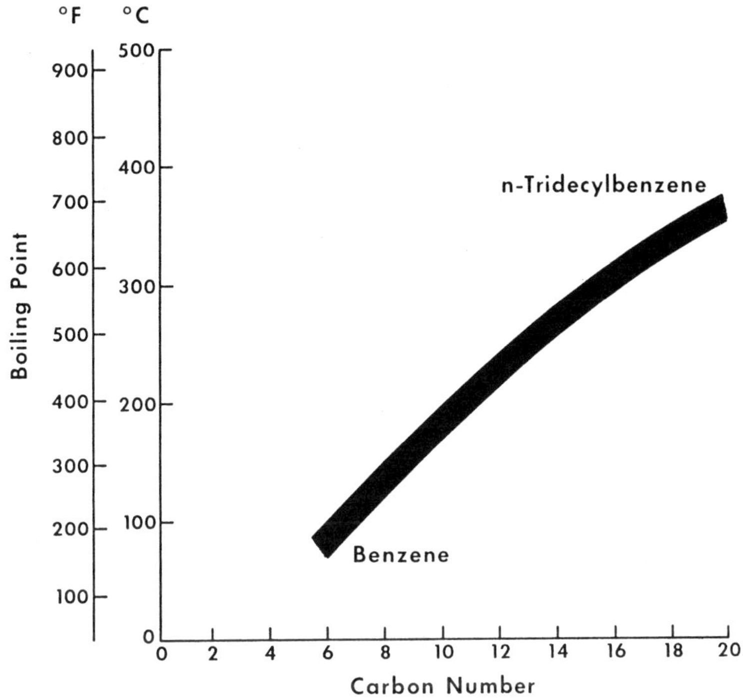

Figure 1-7. Increase in boiling point of an alkyl benzene with increase in carbon number of the alkyl substituent.

Whereas the properties such as viscosity, density, boiling point, and color of petroleum may vary widely among different petroleums, the ultimate or elemental analysis varies, as already noted, over a narrow range for a large number of petroleum samples (Speight, 1991). The carbon content is relatively constant, while the hydrogen and heteroatom contents are responsible for the major differences between petroleums. The nitrogen, oxygen, and sulfur can be present in only trace amounts in some petroleum, which as a result consists primarily of hydrocarbons. On the other hand, a petroleum containing 9.5% heteroatoms may contain essentially no true hydrocarbon constituents insofar as the constituents contain *at least one or more* nitrogen, oxygen, and/or sulfur atoms within the molecular structures.

The molecular species in petroleum vary from simple hydrocarbon species (in the majority) to very complex organic molecules containing atoms of carbon, hydrogen, nitrogen, oxygen, and sulfur (Chapter 4) (in the minority) as well as trace amounts of metals such as vanadium, nickel, iron, and copper (Chapter 3). The complexity of petroleum is further illustrated by the number of potential isomers, i.e., molecules having the same atomic formula, that can exist for a given number of paraffinic carbon atoms, which increases rapidly as molecular weight increases:

Carbon atoms per hydrocarbon	Number of isomers
4	2
8	18
12	355
18	60,523

This same increase in number of isomers with molecular weight also applies to the other molecular types present. Since the molecular weights of the molecules found in petroleum can vary from that of methane (CH_4; molecular weight = 16) to several thousand (Speight et al., 1985), it is clear that the heavier nonvolatile fractions can contain virtually unlimited numbers of molecules. However, in reality the number of molecules in any specified fraction is limited by the nature of the precursors of petroleum as well as by the conditions prevalent during the maturation (conversion of the precursors) processes.

Nevertheless, whereas it is relatively easy to analyze lower molecular weight fractions of petroleum such as gasoline, the analysis of the nonvolatile higher molecular weight (often referred to as *heavy*) fractions such as residua and asphalt is beyond present analytical capability. For these latter fractions, techniques have been developed to separate the sample into concentrates of different molecular types. The properties and amounts of these concentrates are then used to explain or predict the properties of the original sample.

The foregoing comments lead to the conclusion that a detailed analysis may apply to a specific petroleum sample. However, the generic name *petroleum* describes more accurately what *could be* in a sample of petroleum rather than what is *actually* in it.

5.0 COMPOSITION

Petroleum is a very complex mixture consisting of paraffin, naphthene (*cycloparaffin*), and aromatic hydrocarbons as well as nitrogen-, oxygen-, sulfur-containing compounds and traces of a variety of metal-containing compounds, and the amounts of nonhydrocarbon compounds increase with boiling point and, by inference, with molecular weight (Figure 1-8).

There are many studies that have been applied as a means of evaluating petroleum composition. In fact, these studies are so numerous as to be the subject matter of several books and review articles (Rostler, 1965; Altgelt and Gouw, 1979; Speight, 1991). These will not be repeated here, as it is the purpose of this chapter to review methods that have been developed as standard methods for the study of petroleum composition as well as some methods that have seen frequent use or offer a different perspective to composition studies.

Although there are thousands of fractionation techniques, many are variations on the more standard methods. On the other hand, data from some of the methods may be difficult to compare because of subtle differences in applying the technique from one laboratory to another. Finally, the methods reviewed here are those that are applied to the separation of petroleum into three to five major fractions from which refining behavior can be determined. A particular technique that promotes the fractionation of petroleum into a dozen, or often more, fractions may have limited additional use in terms of predicting the gross behavior of petroleum during refinery operations. It is the former endeavors, rather than the latter, that are addressed in this chapter.

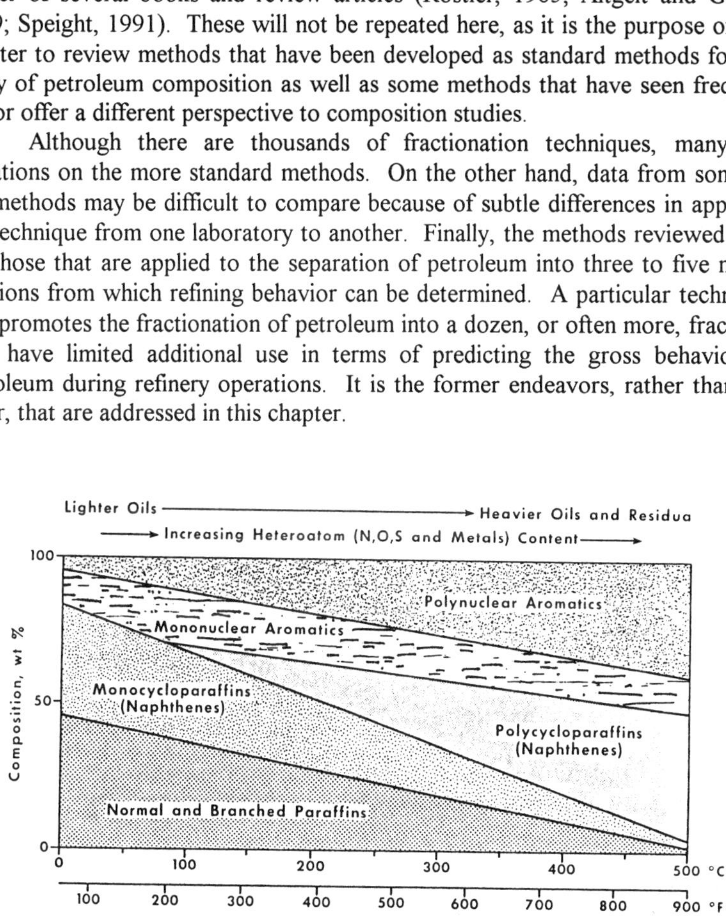

Figure 1-8. Fractionation of petroleum showing the increase in condensed ring systems with boiling point.

In the early days of petroleum processing, it was recognized that the wide molecular weight range of petroleum made it less satisfactory than kerosene as a lamp oil. Because distillation was already well known from the wine industry, it became the process of choice both for molecular weight analysis and for producing high-quality refined products. Before the days of modern instrumental analysis, the laboratory analytical techniques and the refining processes developed together from the same concepts. For this reason, the technique for separation by molecular weight in both the laboratory and the refinery was distillation followed by solvent precipitation for nonvolatile high molecular weight materials.

As the character of the petroleum feedstocks changed and it was realized that petroleum does vary in composition, making the industry more competitive, product quality concerns became more important. It became evident that the heteroatom constituents, namely the nitrogen-, oxygen-, and sulfur-containing constituents as well as the metalloporphyrins (with vanadium, nickel, or iron as the principal metal), were present at only trace levels in the lower boiling fractions of petroleum. However, the quantities of these heteroatoms found in higher boiling fractions were interpreted to mean that many of the molecules contained at least one heteroatom and perhaps more than a single heteroatom.

The early technique used to analyze for molecular type was solid-liquid chromatography using a solid adsorbent such as clay, alumina, or silica gel (Altgelt and Gouw, 1979). The sample was adsorbed on the solid from a suitable carrier solvent, and the different molecular types were then desorbed successively by a series of eluting solvents having increased attraction for the solid adsorbent. The sample was then recovered by evaporating off the elution solvent.

Because of the complexity of petroleum with its wide distributions of molecular weights and molecular types, there has been a continuing effort to divide these distributions into a few standard ranges. For molecular weight, it was done by using boiling point and the typical boiling point ranges of the major refined products. For the molecular types, the fractions were identified by how strong an eluting solvent was required to remove the fraction from the solid adsorbent. Using these techniques together gave fractionation of both distributions.

5.1 Separation by Molecular Weight

The original method of petroleum separation involved molecular weight separation by distillation. Although this technique has been practiced in the laboratory, its application to the heavier crude oils is somewhat limited, but it remains a major operation in refineries (Figure 1-9) and is described here in this context.

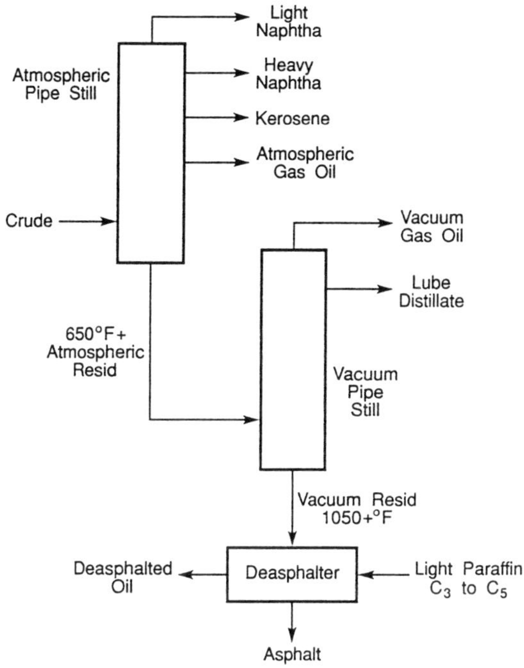

Figure 1-9. Refinery distillation.

Atmospheric and vacuum pipe stills are distillation units that contain a large number of theoretical plates, and the side streams are taken off at different heights up the distillation tower. The petroleum is partially or totally vaporized in a furnace and then fed to the bottom of the atmospheric still. The distillation tower separates the petroleum by boiling point, i.e., molecular weight, with the lighter molecules concentrating at the top of the column and the heavier ones at the bottom. Naphtha is the precursor of gasoline and boils over the same range, while the atmospheric residuum has an initial boiling point on the order of 350°C (660°F). The atmospheric residuum can be distilled further in a vacuum distillation unit to produce overhead fractions such as vacuum gas oils and lubricating oil distillates, and the *bottoms* is a vacuum residuum with an initial boiling point of about 1050°F (565°C.). Finally, the vacuum residuum is deasphalted by mixing it with a light paraffin hydrocarbon, usually propane, to produce an insoluble asphalt and a soluble deasphalted oil.

The amounts of the various fractions obtained from a given petroleum by distillation depend on the nature of the crude oil, which is reflected in the boiling point profile (Figure 1-10). Clearly, petroleum character is very important to the refiner in meeting the product demands of customers as well as in determining what further processing the distillation products may require. For example, the distillates from Louisiana crude oil are of high quality and are suitable feedstocks for lubricating oils and specialty products. On the other hand, the distillates from Bachaquero crude oil are destined primarily for fuels. Thus, the boiling range of petroleum along with its quality is so important to a refiner in meeting the product slate that most refineries must be able to meet any changes in feedstock quality by adapting the refining options. In order do this, the quality of the feedstock must be understood.

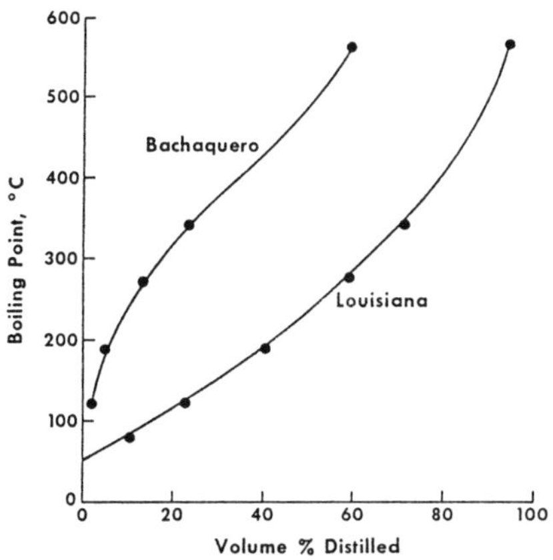

Properties of the 565°C⁺ Residua:

	Louisiana	Bachaquero
Gravity, API	13.1	2.8
Sulfur, wt. %	0.9	3.71
Nitrogen, wt. %	0.4	0.6
Con. Carbon, wt. %	15.8	27.5
Nickel, ppm	20	100
Vanadium, ppm	8	900
Pour Point, °C		55

Figure 1-10. Boiling point profiles for two crude oils and the properties of their respective 565°C$^+$ (1050°F$^+$) residua.

Solvent methods have also been applied to petroleum fractionation on the basis of molecular weight. The major molecular weight separation process used in the laboratory as well as in the refinery is solvent precipitation, which occurs in a refinery in a deasphalter (Figure 1-9) and is essentially an extension of the procedure for separation by molecuar weight, although some separation by polarity might also be operative. The deasphalting process is usually applied to the higher molecular weight fractions of petroleum, such as atmospheric and vacuum residua.

In the laboratory, *n*-pentane and *n*-heptane are the solvents of choice (although other solvents can be used; Speight, 1979) and cause the separation of asphaltenes as brown-to-black powdery materials. In the refinery, supercritical low molecular weight hydrocarbons (e.g., liquid propane, liquid butane, or mixtures of both) are the solvents of choice and the product is a semisolid (tacky) to solid asphalt. The amount of asphalt that settles out of the paraffin/residuum mixture depends on the size of the paraffin, the temperature, and the paraffin-to-feedstock ratio (Girdler, 1965; Corbett and Petrossi, 1978; Speight et al., 1984) (Figure 1-11).

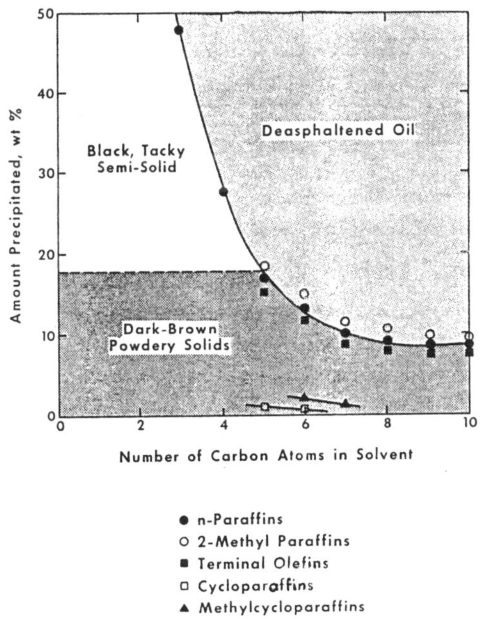

Figure 1-11. Phase diagram for the deasphalting procedure.

In general, residua are considered to be colloidal solutions of high molecular weight asphaltenes stabilized by lower molecular weight molecules of the same general structure (Chapter 4). Addition of a low-boiling hydrocarbon such as pentane, hexane, or heptane causes the asphaltenes to be stripped of the lower molecular weight components and the asphaltenes separate as a solid phase (precipitate). On the other hand, paraffins such as propane and butane (which are gaseous under normal conditions of temperature and pressure) are less effective solvents and cause both the asphaltenes and some of the lower molecular weight polar compounds to separate.

With the wide distribution of molecular weight species in a residuum and in the corresponding asphaltenes (Figure 1-12), the reproducibility of the procedure might be questioned. In response to such a question, when asphaltenes are precipitated (using *n*-heptane) from atmospheric and vacuum residua derived from a different crude oil, the molecular weight distributions for these asphaltenes (Figure 1-13) are essentially identical, even though the residua from which they were derived had very different boiling ranges, indicating that the precipitation (using *n*-heptane) of asphaltenes is quite reproducible.

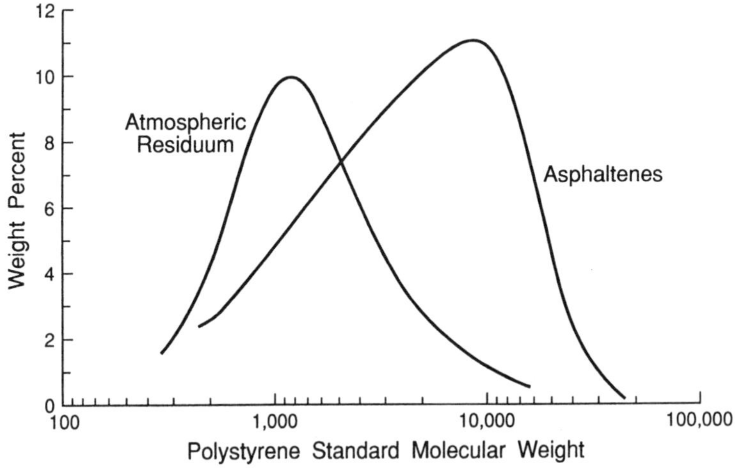

Figure 1-12. Molecular weight distribution in a crude oil and in the related residuum.

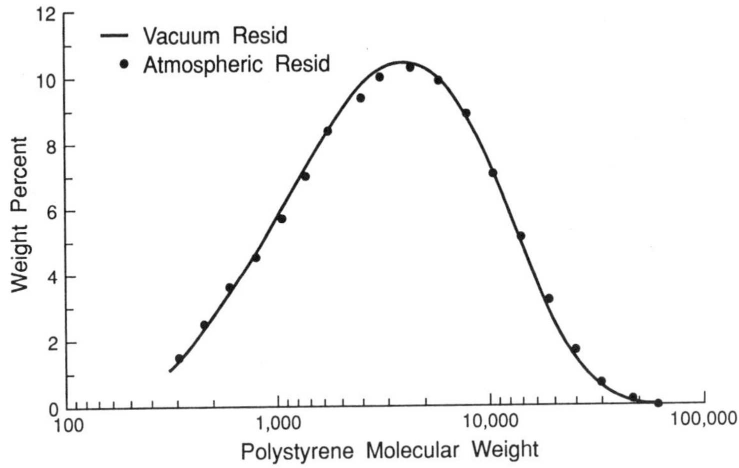

Figure 1-13. Molecular weight distribution of asphaltenes from a crude oil and in the related residuum (c.f., Figure 12).

However, there is evidence that during the exposure of crude oil to the temperatures (approximately 338°C, 640°F) required for distillation, some changes in the functional group distribution might be observed (Francisco and Speight, 1990). The major effect of distillation is the concentration of the nitrogen-containing species into the residuum (Long and Speight, 1990). This is reflected in the concentration into the residuum of the metal-containing constituents as well as those constituents that have a high propensity for the formation of thermal coke (Figure 1-14). In addition, the decarboxylation of carboxylic acid functions occurs, but there does not appear to be any major rearrangement of the carbon skeleton.

5.2 Separation by Molecular Type

Laboratory separations by molecular weight, using distillation or solvents, are helpful in predicting the outcome of refining operations from a product yield standpoint. However, they generate only limited information about the molecular types in the feedstock (mostly by inference) and, thus, about product quality.

Molecular types can be separated into concentrates of given molecular types by various adsorption techniques. These concentrates are then analyzed, both chemically and spectroscopically, to determine in detail what molecular types are in the concentrates. This adsorption separation can divide wide-boiling fractions into molecular type concentrates. However, the separations are somewhat sharper if the feed is divided into narrower molecular weight ranges before the adsorption separation.

Many analytical fractionation studies have been applied to petroleum as a means of evaluating composition and as a guide to feedstock behavior in various processes (Corbett and Petrossi, 1978; Altgelt and Gouw, 1979; Tissot, 1984; Speight, 1991).

The terminology of the various methods might differ but, in general terms, group-type analysis of petroleum is often identified by the acronyms for the names: PONA (paraffins, olefins, naphthenes, and aromatics), PIONA (paraffins, isoparaffins, olefins, naphthenes, and aromatics), PNA (paraffins, naphthenes, and aromatics), PINA (paraffins, isoparaffins, naphthenes, and aromatics), and SARA (saturates, aromatics, resins, and asphaltenes). However, it must be recognized

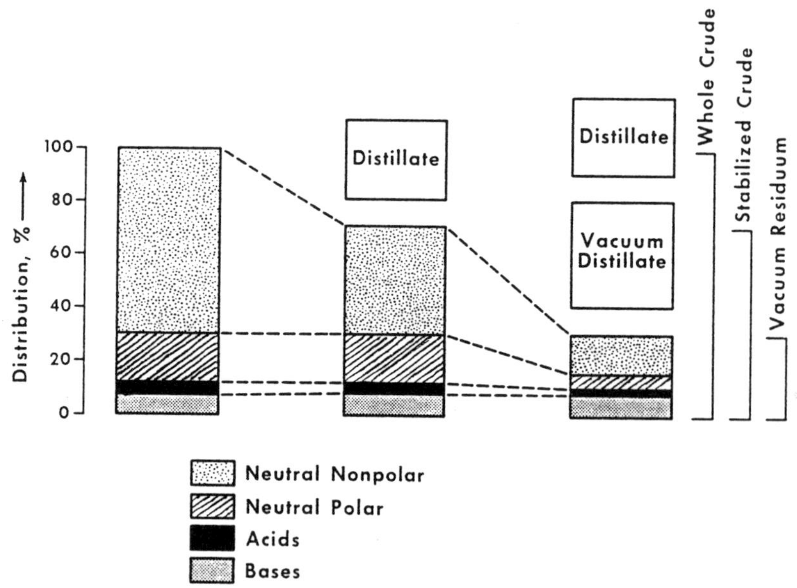

Figure 1-14. Concentration of polar constituents during crude oil distillation.

that the fractions produced by the use of different adsorbents will differ in content and will also be different from fractions produced by solvent separation techniques.

Two analytical procedures have received considerable attention over the years: (1) the United States Bureau of Mines–American Petroleum Institute (USBM-API) method (McKay et al., 1977; Francisco and Speight, 1984) and (2) the saturates-aromatics-resins-asphaltenes (SARA) method (Jewell et al., 1974). This latter method, often also called the saturates-aromatics-polars-asphaltenes (SAPA) method, involves the prior separation of asphaltenes before contact of the feedstock with the adsorbents, and it is preferred by many investigators because of the potential for irreversible adsorption of the asphaltenes onto many adsorbents. Compounds can be isolated and identified from complex matrices, while detailed *analytical fingerprinting* of petroleum constituents and petroleum products is feasible (Altgelt and Gouw, 1975).

A simple alternative to the SARA sequence is the separation of a deasphaltened oil on deactivated silica or alumina with heptane followed by elution with benzene and then with benzene/methanol (Speight, 1991. Another example of a similar procedure involves fractionation into four gross fractions and subfractionation of these four fractions (Boyd and Montgomery, 1963). It has also been reported (Oudin, 1970) that chromatography using a combination of alumina and silica gel is suitable for deasphaltened feedstocks or products that do not contain asphaltenes. There has also been a report (Al-Kashab et al., 1976) of the direct fractionation of hydrocarbon and heterocompounds from deasphaltened residua on a dual alumina-silica column with subsequent treatment of the polar fraction with cation and anion exchange resins into basic, acidic, and neutral materials. The method also includes chromatography of the asphaltenes, but only with highly polar asphaltene samples are the basic and acidic compounds first removed with ion exchange resins. The remainder of the feedstock is separated into saturates, aromatics, and heterocompounds using alumina-silica adsorption. Separation of the saturates into *n*-alkanes and isoalkanes plus cycloalkanes is achieved by use of urea and thiourea.

It is, of course, worth noting at this point that there are, indeed, American Society for Testing and Materials (ASTM) *standard methods* for the separation of petroleum into constituent fractions. In these methods (American Society for Testing and Materials, 1970, 1980, 1984), not only are the procedures specified in great detail but so are the materials and the equipment employed. In keeping with their goals, the ASTM require reproducibility and precision. Three methods essentially provide for the separation of a feedstock into four or five constituent fractions (Figures 1-18, 1-19, and 1-20). It is interesting to note that as the methods have evolved there has been a change from the use of pentane (ASTM D-2006 and ASTM D-2007) to heptane (ASTM D-4124) to separate asphaltenes.

This is, in fact, in keeping with the production of a more consistent fraction that represents these higher molecular weight, more complex constituents of petroleum.

Two of the methods (ASTM D-2007 and ASTM D-4124) use adsorbents to fractionate the deasphaltened oil, but the third method (ASTM D-2006) advocates the use of various grades of sulfuric acid to separate the material into compound types. Caution is advised in the application of this method, because in the present author's experience the method works well with some feedstocks but not with *all* feedstocks.

A note of caution must be added here about any method in which the asphaltenes are first isolated as a separate fraction. For many years, the method of asphaltene separation was not standardized, and even now it remains subject to the preferences of the standards orgainzations of different countries. The use of both pentane and heptane has been widely advocated, and although heptane is becoming the deasphalting liquid of choice, this is by no means a hard-and-fast rule. And it must be recognized that large volumes of solvent may be required to effect a reproducible separation, similar to amounts required for consistent asphaltene separation (IP, 1990; Speight, 1991; ASTM D-2006; ASTM D-2007; ASTM D-4124; ASTM D-893). It is also preferable that the solvents be of sufficiently low boiling point that complete removal of the solvent from the fraction can be effected and, most important, the solvent must not react with the feedstock. Hence, the preference for hydrocarbon liquids, although the several standard methods that have been used are not unanimous in the ratio of hydrocarbon liquid to feedstock.

Method	Deasphalting liquid	Volume ml/g
ASTM D-893	*n*-pentane	10
ASTM D-2006	*n*-pentane	50
ASTM D-2007	*n*-pentane	10
IP 143	*n*-heptane	30
ASTM D-3279	*n*-heptane	100
ASTM D-4124	*n*-heptane	100

However, it must be recognized that some of these methods were developed for use with feedstocks other than heavy oil and adjustments are necessary.

Asphaltene yield varies with the hydrocarbon used for the separation and with other factors such as the ratio of the deasphalting liquid to the sample as well as the contact time (Speight et al., 1984). Leaving asphaltene material in the adsorption column feedstock can give rise to problems in which the asphaltene material in the deasphaltened oil undergoes irreversible adsorption on the solid adsorbent.

One of the common problems of any of the fractionation schemes just described is the nature of the adsorbent. In the early reports on petroleum fractionation, *clay* often appeared as an adsorbent to effect separation of the feedstock into various constituent fractions. However, *clay* (*fuller's earth, attapulgus clay attapulgite*, and the like) is often difficult to define with any degree of precision from one batch to another. Variations in the nature and properties of the clay can, and will, cause differences not only in the yields of composite fractions but also in the distribution of the compound types in those fractions. In addition, irreversible adsorption of the more polar constituents on the clay can be a serious problem when further investigations of the constituent fractions are planned.

One option for resolving this problem has been to use more *standard adsorbents* such as alumina and silica. These materials are more easy to define and are often accompanied by guarantees of composition and type by various manufacturers. They also tend to adsorb adsorb less of the feedstock than a clay. Once the nature of the adsorbent is guaranteed, reproducibility becomes a reality. Without reproducibility the analytical method does not have credibility.

Gel permeation chromatography is an attractive technique for the determination of the number average molecular weight (M_n) distribution of petroleum fractions, especially the heavier constituents, and petroleum products (Altgelt, 1968; Oelert, 1969; Altgelt, 1970; Baltus and Anderson, 1984; Hausler and Carlson, 1985; Reynolds and Biggs, 1988).

Ion exchange chromatography is also widely used in the characterization of petroleum constituents and products. For example, cation exchange chromatography can be used primarily to isolate the nitrogen constituents in petroleum (Snyder and Buell 1965; Drushel and Sommers, 1966; McKay et al., 1974), thereby giving an indication of how the feedstock might behave during refining and also an indication of any potential deleterious effects on catalysts.

Liquid chromatography has helped to characterize the group composition of crude oils and hydrocarbon products since the beginning of this century. The type and relative amount of certain hydrocarbon classes in the matrix can have a profound effect on the quality and performance of the hydrocarbon product. The *fluorescent indicator adsorption* (*FIA*) method (ASTM D-1319) has been used to measure the paraffinic, olefinic, and aromatic content of gasoline, jet fuel, and liquid products in general (Suatoni and Garber, 1975; Miller et al., 1983; Norris and Rawdon, 1984).

High-performance liquid chromatography (*HPLC*) has found great utility in separating different hydrocarbon group types and identifying specific constituent types (Colin and Vion, 1983; Drushel, 1983; Miller et al., 1983; Chartier et al., 1986). Of particular interest is the application of the HPLC technique to the identification of the molecular types in the heavier feedstocks, especially the molecular types in the asphaltene fraction, which may be especially useful for studying such materials on a *before-processing* and *after-processing* basis (Speight, 1986), and the products of their processing (Chmielowiec et al., 1980 Alfredson, 1981; Bollet et al., 1981; Colin and Vion, 1983; George and Beshai, 1983; Felix et al., 1985; Coulombe and Sawatzky, 1986; Speight, 1986; Speight, 1991).

Several recent high-performance liquid chromatographic separation schemes are applicable because they also incorporate detectors not usually associated with conventional hydrocarbon group-type analyses (Matsushita et al., 1981; Miller et al., 1983; Rawdon, 1984; Lundanes and Greibrokk, 1985; Hayes and Anderson, 1985; Schwartz and Brownlee, 1986; Hayes and Anderson, 1987).

The general advantages of high-performance liquid chromatography are that (1) each sample may be analyzed *as received* even though the boiling range may vary over a considerable range; (2) the total time per analysis is usually on the order of minutes; and, perhaps most important, (3) the method can be adapted for on-stream analysis in a refinery.

In recent years, *supercritical fluid chromatography* has found use in the characterization and identification of petroleum constituents and products. A supercritical fluid is defined as a substance above its critical temperature. A primary advantage of chromatography using supercritical mobile phases results from the mass transfer characteristics of the solute. The increased diffusion coefficients of supercritical fluids compared with liquids can lead to greater speed in separations or greater resolution in complex mixture analyses. Another advantage of supercritical fluids compared with gases is that they can solubilize thermally labile and nonvolatile solutes and, upon expansion (decompression) of this solution, introduce the solute into the vapor phase for detection (Lundanes et al., 1986; Roberts, 1995).

Currently, supercritical fluid chromatography is leaving the stages of infancy (Smith et al., 1988). The indications are that it will find wide applicability to the problems of characterization and identification of the higher molecular weight species in petroleum (Schwartz et al., 1988), thereby adding an extra dimension to our understanding of refining chemistry. It will still remain an option as a means of product characterization, although the use may be somewhat limited because of the ready availability of other characterization techniques.

One of the common disadvantages of feedstock and product evaluation by the use of adsorbent fractionation is the ill-defined nature of the adsorbent. Thus, (1) performance will, and often does, differ with the same feedstock/product and

(2) chemical and physical modification of the constituents can occur through excessive heat of adsorption. One option for resolving these issues has been the use of more "standard" adsorbents such as alumina and silica. The composition of these materials is easier to guarantee in terms of composition and type by various manufacturers. Once the nature of the adsorbent is guaranteed, reproducibility becomes a reality. Without reproducibility any analytical method does not have credibility.

The use of analytical *fractionation by chemical agents* for feedstocks and products is also fraught with difficulties. For example, the use of sulfuric acid should be advocated only with caution. Many feedstocks and products react with the acid to such an extent that irreversible chemical change and/or emulsion formation occurs. The use of chemical agents for feedstock and product characterization is somewhat more difficult to combat than the use of adsorbents. Chemical reactions occur during the procedure, making alternative routes difficult to conceive.

A suggested alternative suggestion to the use of solid adsorbents or of chemical agents involves the implementation of all-solvent procedures (Rostler, 1965; Speight, 1979). Again, as with adsorbent procedures, the names of the fractions are operational definitions only and are not guarantees of a degree of compositional purity of the fractions. The obvious benefit of such a technique is the complete recovery of material, thereby allowing a complete characterization of the feedstocks and products.

6.0 MOLECULAR TYPES IN PETROLEUM

As already stated, petroleum is a complex mixture of paraffin, naphthene, and aromatic hydrocarbons as well as nitrogen-, oxygen-, and sulfur-containing compounds and traces of a variety of metal-containing compounds, and the amounts of nonhydrocarbon compounds increase with molecular weight. The distribution of the constituents of petroleum varies with the petroleum but can generally be represented as consisting of the various hydrocarbon species with varying amounts of heteroatom species (Figure 1-8).

By definition, the *saturates* consist of paraffins and cycloparaffins (naphthenes). The single-ring *naphthenes*, or *cycloparaffins*, present in petroleum are primarily alkyl-substituted cyclopentane and cyclohexane. The alkyl groups are usually quite short, with methyl, ethyl, and isopropyl groups the predominant substituents. As the molecular weight of the naphthenes increases, the naphthene fraction contains more condensed rings with six-membered rings predominating. However, five-membered rings are still present in the complex higher molecular weight molecules. The *aromatics* vary from *monoaromatics* (containing one benzene ring in a molecule) to *diaromatics* (substituted naphthalene) to *triaromatics* (substituted phenanthrene). Higher condensed ring systems (*tetraaromatics, pentaaromatics*) are also known but are somewhat less prevalent than the lower ring systems, and each aromatic type will have increasing amounts of condensed ring naphthene attached to the aromatic ring as molecular weight is increased.

The typical *nitrogen compounds* found in petroleum are generally divided into two groups, basic and nonbasic (Figure 1-15), each of which has alkyl chains and other ring systems. The basic nitrogen compounds cause difficulty with the many acid-catalyzed processes used in petroleum refining. For example, in catalytic cracking, the basic nitrogen adsorbs on the catalytic acid sites and reduces the cracking activity of the catalyst. Furthermore, in reactions catalyzed by liquid acids, the presence of basic nitrogen compounds in the feed increases acid

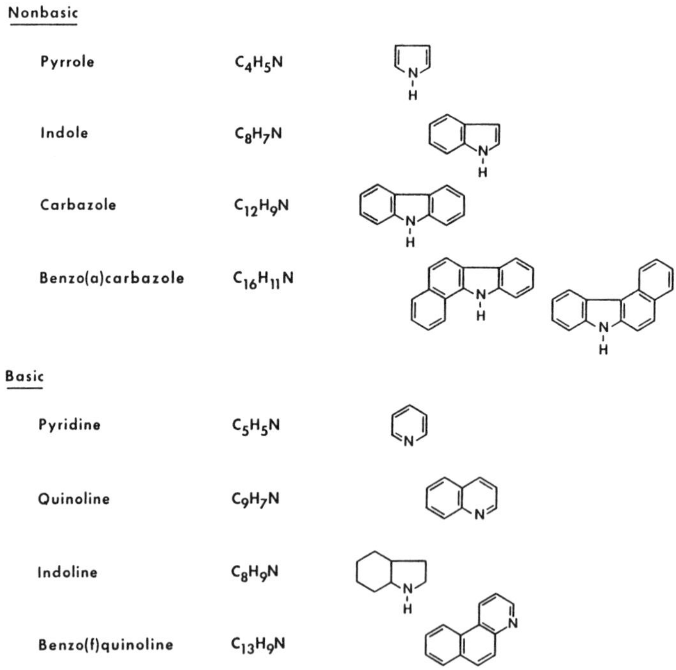

Figure 1-15. Subdivision of oranonitrogen compounds as nonbasic and basic.

consumption and, thus, the cost of the process. Typical of such processes are alkylation, isomerization, and olefin absorption.

Another type of nitrogen compound is the *porphyrins*, which consist of four pyrrole rings connected together with methylene bridges at the carbons next to the nitrogen atoms. They are normally found in trace quantities in the high molecular weight fractions of petroleum as metal complexes (Chapter 3).

The *oxygen compounds* found in petroleum fractions are often products of exposure to air. However, some legitimate oxygenated compounds do exist in petroleum; these are typically phenols, naphthenic acids, and esters. They increase in quantity and complexity as molecular weight increases, just as sulfur and nitrogen compounds do.

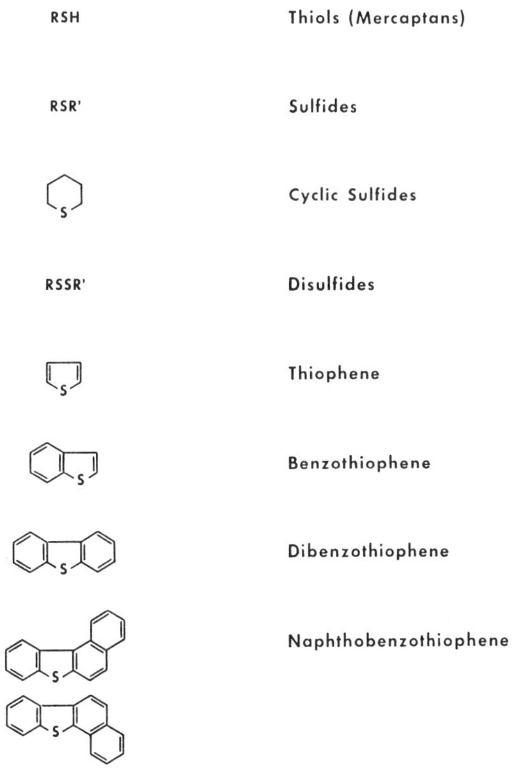

Figure 1-16. Types of sulfur compounds in petroleum.

A variety of *sulfur compounds* occur in petroleum (Figure 1-16) and include mercaptans (–SH), sulfides (–S–), and disulfides (–S–S–). They are much less thermally stable than the thiophene derivatives and often lose hydrogen sulfide on heating. They can also react thermally to form more stable sulfur compounds. In the chromatographic separation, these sulfur compounds are found in the polar aromatics fraction, even though they may not all be aromatic. *Thiophenes* tend to exhibit aromatic behavior and are collected with the aromatics when adsorption separation is used on petroleum fractions.

7.0 COMPOSITION MAPS

Petroleum, being classed as a hydrocarbon resource, is composed of a variety of hydrocarbon molecular species (Figure 1-8). For an individual crude oil, it is the relative proportions of the individual molecular species that dictate the character of the crude oil.

However, the composition of petroleum can be viewed as consisting of two continuous distributions, one of molecular weight and the other of molecular type. Early attempts to study the composition of petroleum involved subdivision into fractions conveniently named *saturates*, *aromatics*, and *polar aromatics* (*resins*) by adsorption chromatography after a preseparation of the *asphaltenes* by use of a low-boiling liquid hydrocarbon (Figure 1-8). Thus, the early composition maps were plots of yield versus molecular type with the data obtained by running clay-gel separations on the molecular weight fractions obtained by distillation and solvent precipitation (Speight, 1991).

More assiduous fractionation of petroleum showed the variation of heteroatom content with the various fraction, and more recent work allowed petroleum to be mapped on the basis of boiling point and carbon number. More detailed maps evolved in which, using the molecular weights and elemental analyses, the number of nitrogen and sulfur atoms in the aromatic and polar aromatic fractions were also exhibited (Speight, 1991). These data are useful insofar as they show that not only can every molecule in the resins and asphaltenes have more than one sulfur or nitrogen but also some molecules probably contain both sulfur and nitrogen. As the molecular weight of the aromatic fractions decreases, the sulfur and nitrogen contents of the fractions also decrease. In contrast to the sulfur-containing molecules, which appear in both the naphthene-aromatics and the polar aromatics fractions, the oxygen compounds present in the heavy fractions of petroleum are normally found in the polar aromatics fraction.

The data also show the variation in composition of the different fractions as the molecular weight of the fractions is increased. The saturate fraction disappears as the molecular weight approaches that of asphaltenes. On the other hand, it tends to increase as molecular weight is reduced to that of distillates boiling below the resid. The naphthene-aromatics tend to remain fairly constant throughout the boiling range of the residuum, while the polar aromatics increase as molecular weight increases. The polar aromatics decrease as molecular weight is reduced below that of the residuum fractions and are essentially zero in the light distillates of the petroleum.

Other data show that the saturates are most plentiful in the lower boiling fractions and essentially disappear by the time the resins and asphaltenes are reached. The aromatics are essentially constant through the volatile fractions but also are gone by the time the resins and asphaltenes are reached. Polar aromatics increase with molecular weight and account for the entire fraction for resins and asphaltenes. Thus, the resin and asphaltene fractions do not contain any hydrocarbons in this resid sample. It must be remembered that this type of map is specific to the feedstock used and will vary with both the petroleum source and the boiling range of the feedstock sample.

More recent work (Long and Speight, 1989) involved the development of a different type of compositional map for various petroleums using the molecular weight distribution and the molecular type distribution as coordinates. The separation involved the use of an adsorbent such as clay, and the fractions were characterized by solubility parameter as a measure of the polarity of the molecular types. The molecular weight distribution can be determined by gel permeation chromatography. Using these two distributions, a map of composition can be prepared using molecular weight and solubility parameter as the coordinates for plotting the two distributions. Such a composition map can provide insights into many separation and conversion processes used in petroleum refining.

The molecular type was characterized by the polarity of the molecules, as measured by the increasing adsorption strength on an adsorbent. At the time of the original concept, it was unclear how to characterize the continuum in molecular type or polarity. For this reason, the molecular type coordinate of the first maps was the yield of the molecular types ranked in order of increasing polarity. However, this type of map can be somewhat misleading, because the areas are not related to the amounts of material of a given type. The horizontal distance on the plot is a measure of the yield and there is not a continuous variation in polarity for the horizontal coordinate. It was suggested that the solubility parameter of the different fractions could be used to characterize both polarity and adsorption strength.

In order to attempt to remove some of these potential ambiguities, more recent developments of this concept have focused on the solubility parameter. The simplest map that can be derived using the solubility parameter is produced with the solubility parameters of the solvents used in solvent separation procedures, equating these parameters to the various fractions (Figure 1-17).

However, the solubility parameter boundaries determined by the values for the eluting solvents that remove the fractions from the adsorbent offer a further step in the evolution of petroleum maps. Maps were obtained using three solvents, cyclohexane, toluene, and 5% water in tetrahydrofuran, that are typical solvents for chromatographic separation of saturates, aromatics, and polar species. Elution from the adsorbent using more complex solvent sequences yields a somewhat more detailed map in terms of the solubility parameter relationships of the various fractions. Furthermore, the solubility parameters of the individual eluted fractions in the eight-solvent case have also been measured and agree with these values, and the method gives an effective continuum in polarity for the horizontal scale.

Measuring the overall solubility parameter of a petroleum fraction is a time-consuming chore. Therefore, it is desirable to have a simpler, less time-

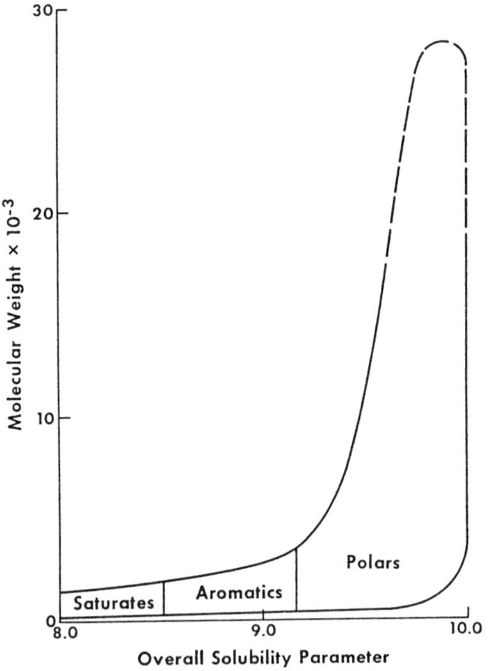

Figure 1-17. A petroleum composition map based on molecular weight and solubility parameter.

consuming measurement that can be made on petroleum fractions which will correlate with the solubility parameter and thus give an alternative continuum in polarity. In fact, the hydrogen-to-carbon atomic ratio and other properties of petroleum fractions that can be correlated with the solubility parameter (Wiehe, 1992; Speight, 1994; Wiehe, 1995; Wiehe, 1996) also provide correlations for the behavior of crude oils.

7.1 Predicting Properties from a Composition Map

Although there are many maps that can be derived from the fractionation of petroleum, for convenience only selected maps will be used. Thus, a composition map can be used to show where a particular physical or chemical property tends to concentrate on the map (Figure 1-18).

The coke-forming propensity, i.e., the amount of the carbon residue, is shown for various regions on the map for a sample of atmospheric residuum. The plot shows molecular weight plotted against weight percent yield in order of increasing polarity. The dashed line is the envelope of composition of the total sample. The slanted lines show the boundaries of solvent-precipitated fractions, and the vertical lines show the boundaries of the fractions obtained by clay adsorption of the pentane-deasphalted oil.

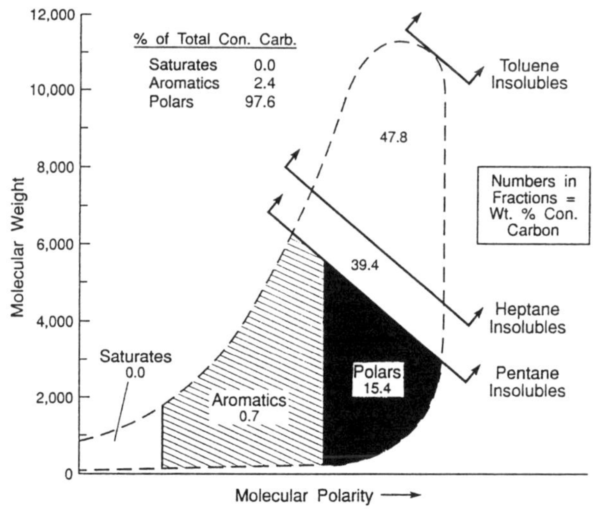

Figure 1-18. Property predicition using a composition map.

The boundary for toluene precipitation shows that there are no toluene-insoluble materials in this particular resid because there is no material of sufficiently high polarity or molecular weight to be insoluble in toluene. If there were, it would have a carbon residue greater than 50. Moving down the right-hand side of the map, the first fraction is the *n*-heptane-insoluble fraction or heptane asphaltenes. The carbon residue of this fraction is 47.8, which is typical of the value of about 50 for heptane asphaltenes from a wide variety of petroleum sources. Moving down to the pentane asphaltene boundary, an insoluble fraction is shown that includes all the heptane asphaltenes as well as lower molecular weight polar species and a small amount of high molecular weight aromatics. This encompasses an insoluble fraction with a carbon residue of 39.4.

The region of the map below the pentane-insoluble boundary corresponds to pentane-deasphalted oil from the original residuum. The saturate, aromatic, and polar fractions were separated by adsorption of the deasphalted oil over clay. The saturates show a zero carbon residue and the aromatics are only a little higher at 0.7%. The coke-forming constituents in the deasphaltened oil are the polar aromatics, which have a carbon residue of 15.4. The carbon residue balance shown in the insert table shows that almost all of the coke-forming materials (97.6%) are in the asphalt and other lower molecular weight polar compounds. This is further evidence of the similarity, with the exception of molecular weight, between asphaltenes and the lower molecular weight polar species in deasphalted oil.

7.2 Predicting Separations from a Composition Map

A composition map can be very useful for predicting the effectiveness of various types of separations or conversions of petroleum (Figure 1-19). These processes are adsorption, distillation, solvent precipitation with relatively nonpolar solvents, and solvent extraction with polar solvents. The vertical lines show the cut points between saturates aromatics and polar aromatics as determined by clay chromatography. The slanted lines show how distillation, extraction, and solvent precipitation can divide the composition map. The line for distillation divides the map into distillate, which lies below the dividing line, and bottoms, which lies above the line. As the boiling point of the distillate is raised, the line moves upward, including higher molecular weight materials and more of the polar species in the distillate and rejecting lower molecular weight materials from the bottoms. As more of the polar species are included in the distillate, the carbon residue of the distillate rises.

The solvent precipitation lines and solvent extraction lines are determined by solubility theory and move up and down depending on the solvent used, as well as the operating conditions of the separation. For example, with *n*-paraffins as precipitating agents, the cut line is approximately that for *n*-heptane deasphaltening. Higher molecular weight *n*-paraffins move the cut line up and lower molecular weight *n*-paraffins move it down. Therefore very hard, high molecular weight asphalt is precipitated by higher paraffins and soft oily asphalt is precipitated by low molecular weight paraffins.

The slope and position of the cut line also depend on the solubility parameter of the precipitating solvent as well as the temperature of operation. For paraffin solvents, the overall solubility parameter of the solvent is to the left of the composition map and the cut line is the right-hand wing of the solubility bell curve as it cuts through the composition map.

The cut line for extraction with polar solvents such as phenol, cresol, *n*-methylpyrrolidone, and furfural is slanted in the opposite direction from solvent precipitation, because the solubility parameter of the polar solvent lies to the right of the composition map. Thus, the cut line is the left-hand wing of the solubility bell curve of the polar solvent. It moves up and down and left and right as the solubility parameter of the polar solvent is changed. Furthermore, it is also sensitive to the operating conditions of the extraction, such as temperature and solvent-to-oil ratio.

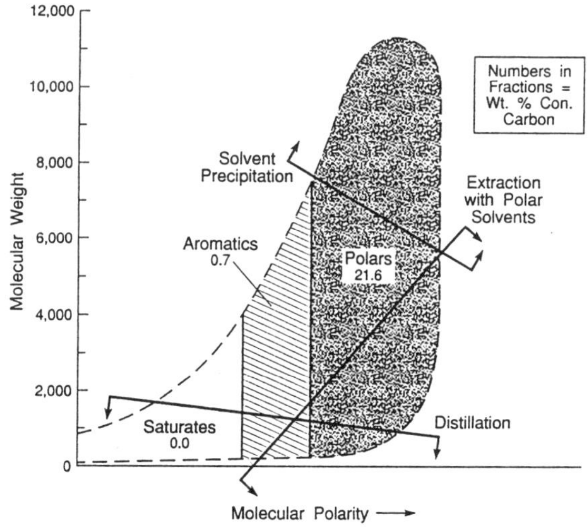

Figure 1-19. Prediction of the effectiveness of various types of separations or conversions using a composition map.

7.3 Predicting Conversions from a Composition Map

In contrast to the cut lines generated by separation processes, conversion processes move materials in the composition from one molecular type to another. For example, reforming converts saturates to aromatics and hydrogenation converts aromatics to saturates and polar aromatics to either aromatics or saturates (Figure 1-20). The processes illustrated are reforming, hydrogenation, hydrotreating, hydrocracking, visbreaking, and heat soaking. The arrows go from the material converted to what it is converted to. Hydrotreating removes nitrogen and sulfur compounds from polar aromatics without much change in molecular weight. Hydrocracking converts polar species to aromatics while at the same time reducing molecular weight. Visbreaking and heat soaking primarily lower or raise the molecular weight of the polar species in the composition map. Thus, visbreaking is used to lower the viscosity of heavy oils, whereas heat soaking is a coking method. Thus, conversion processes can change the shape and size of the composition map.

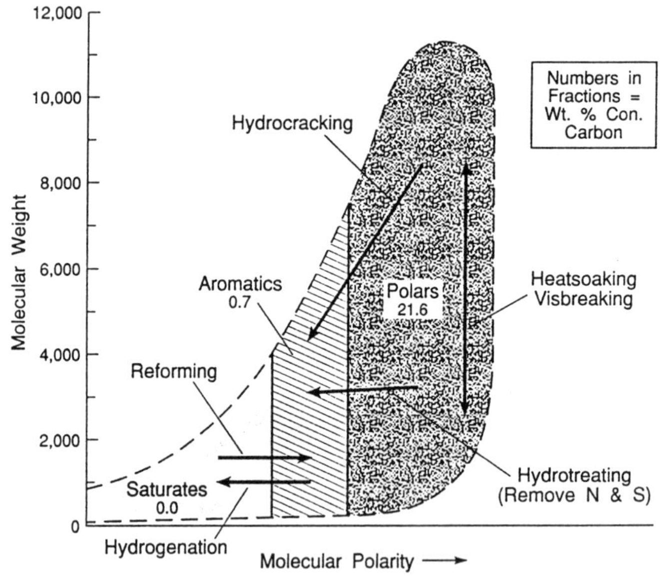

Figure 1-20. Prediction of refining operations using a composition map.

8.0 REFERENCES

Al-Kashab, K., and Neumann, H. J. 1976. Strassen and Tieflau. 30(3): 44.

Altgelt, K. H., and Gouw, T. H. (Editors). 1979. Chromatography in Petroleum Analysis. Marcel Dekker Inc., New York.

American Society for Testing and Materials. 1970. Test No. ASTM D2006. American Society for Testing and Materials, Philadelphia, PA.

American Society for Testing and Materials. 1980 Test No. ASTM D2007. 1980. American Society for Testing and Materials, Philadelphia, PA.

American Society for Testing and Materials. 1984. Test No. ASTM D4124. American Society for Testing and Materials. Philadelphia, PA.

Berkowitz, N., and Speight, J. G. 1975. Fuel. 54: 318.

Bestougeff, M. 1967. In Fundamental Aspects of Petroleum Geochemistry. B. Nagy and U. Colombo (Editors). Elsevier, Amsterdam, The Netherlands. Chapter 3.

Boyd, M. L., and Montgomery, D. S. 1963. J. Inst. Petrol. 49: 345.

Corbett, L. W., and Petrossi, U. 1978. Ind. Eng. Chem. Prod. Res. Dev. 17: 342.

Durand, B. (Editor). 1980. Kerogen: Insoluble Matter from Sedimentary Rocks. Editions Technip, Paris, France.

Francisco, M. A., and Speight, J. G. 1990. Rev. Inst. Fr. Petrole. 45: 733.

Girdler, R. B. 1965. Proc. Assoc. Asphalt Paving Technol. 34: 45.

Hirsch, D. E., Hopkins, R. L., Coleman, H. J., Cotton, F..O., and Thompson, D. J. 1972. Anal. Chem. 44: 915.

Jewell, D. M., Weber, J. H., Bunger, J. W., Plancher, H., and Latham, D. R. 1972a. Anal. Chem. 44: 1391.

Jewell, D. M., Ruberto, R. G., and Davis, B. E. 1972. Anal. Chem. 44: 2318.

Jewell, D. M., Albaugh, E. W., Davis, B. E., and Ruberto, R. G. 1974. Ind. Eng. Chem. Fundam. 13: 278.

Kuuskraa, V. A., and Hammershaimb, E. 1984. Major Tar Sand and Heavy Oil Deposits of the United States. Interstate Oil Compact Commission, Oklahoma City, Oklahoma.

Long, R. B., and Speight, J. G. 1989. Rev. Inst. Francais du Petrole. 44: 205.

Long, R. B., and Speight, J. G. 1990. Rev. Inst. Francais du Petrole. 45: 553.

Oudin, J. L. 1970. Rev. Inst. Fr. Petrole. 25: 470.

Phizackerley, P. H., and Scott, L. O. 1979. In Bitumens, Asphalts and Tar Sands. G. V. Chilingarian and T. F. Yen (Editors). Elsevier, New York, 1979. p. 57.

Roberts, I. 1995. In Chromatography in the Petroleum Industry. E. R. Adlard (Editor). Elsevier, Amsterdam, The Netherlands. Chapter 11.

Rostler, F. S. 1965. In Bituminous Materials: Asphalt, Tars and Pitches. A.

Hoiberg (Editor). Interscience, New York. Vol. II, Part 1, p. 151.

Scouten, C. S. 1990. In Fuel Science and Technoogy Handbook. J. G. Speight (Editor). Marcel Dekker Inc., New York.

Speight, J.G . 1979. Studies on Bitumen Fractionation—(A) Fractionation by a Cryoscopic Method; (B) Effect of Solvent Type on Asphaltene Solubility. Information Series No. 84. Alberta Research Council, Edmonton, Alberta, Canada.

Speight, J. G. 1991. The Chemistry and Technology of Petroleum. 2nd Edition. Marcel Dekker Inc., New York.

Speight, J. G. 1994. In Asphalts and Asphaltenes, 1., T. F. Yen and G. V. Chilingarian (Editors). Elsevier, Amsterdam, The Netherlands. Chapter 2.

Speight, J. G., Long, R. B., and Trowbridge, T. D. 1984. Fuel. 63: 616.

Tissot, B. P., and Welte, D. H. 1978. Petroleum Formation and Occurrence: A New Approach to Oil and Gas Exploration. Springer-Verlag, New York.

Wiehe, I. A. 1992. Ind. Eng. Chem. Res. 31: 530.

Wiehe, I. A. 1995. Ind. Eng. Chem. Res. 34: 661.

Wiehe, I. A. 1996. Fuel Sci. Technol. Int. 14: 289.

Chapter 2

Petroleum Analysis and Evaluation

James G. Speight
Western Research Institute
365 North 9th Street
Laramie
Wyoming 82072

1.0 INTRODUCTION

In the early days of petroleum refining, kerosene became the fraction of choice. The first refineries were essentially distillation units to isolate as much of the kerosene fraction as possible. As the feedstocks to the early stills changed in character, it became necessary to examine chemical and compositional differences between the feedstocks and attempt to find some method of predicting product yields. In addition, the introduction of cracking (thermal decomposition) in some refineries emphasized the need for feedstock and product analysis. Hence, application of analytical techniques, already in use in other industries, became the means of feedstock and product evaluation in the petroleum industry.

In the present context, the *average* quality of petroleum delivered to refineries has diminished by approximately two points on the API gravity scale over the past two decades (Swain, 1991). This has been accompanied by, among other things, an increase in the sulfur content. As a result, refineries have had to adapt processing strategies to deliver liquid fuel products from feedstocks that are deteriorating in quality. As the reserves of lighter crude oils continue to decline, there is an increasing need to develop options for upgrading the abundant supply of known heavy oil reserves into more value-added products. Thus, there is an ever-increasing need to understand the chemical nature of the constituents and the chemistry of conversion processes.

Analytical studies have been intensified to deal with many of the previously unknown feedstock properties that arise as a result of the influx of heavier feedstocks into refineries and to combat the problems that arise during processing. In addition, characterization studies, particularly those related to the higher molecular weight constituents (Speight, 1994a), must be considered an integral part of any analytical methodology.

The continued efforts to analyze and characterize feedstocks frequently lead to a greater understanding the nature of the feedstock. They can also lead to the development of various process options for the conversion of these materials into lighter products.

Petroleum is a naturally occurring material that consists predominantly of hydrocarbons with lesser proportions of constituents that contain nitrogen, oxygen, and sulfur. The latter constituents, often referred to as heterocompounds, are heteroatomic constituents by virtue of the nitrogen and/or oxygen and/or sulfur content. But in the simplest sense, petroleum can be classified as a naturally occurring mixture of hydrocarbons.

It is the purpose of this chapter to present a general overview of the methods used for feedstock characterization as well as methods for product characterization. In addition, some aspects of product distribution will be introduced to illustrate the effects of different feedstocks on the product slate, which will also emphasize the need for characterization techniques.

2.0 DEFINITIONS

The definitions of petroleum and associated materials are varied and often do not give a full or useful description of the material under investigation (Chapter 1). Nevertheless, much of the terminology persisted from the time when it was first introduced to the present.

The petroleum system is complex and is often described as a collection of various chemical types. From an analytical perspective, it can be considered to be a composite of four generic fractions, (1) *saturates*, (2) *aromatics*, (3) *resins*, and (4) *asphaltenes*, which exist in harmonious relationship with each other and can be separated as individual fractions.

It is necessary to stress that these terms are operational definitions only (Chapter 1). They do not accurately describe the true character of the constituents of the fractions, and they are not meant to imply any particular degree of chemical character or of purity. For example, hexylbenzene ($C_6H_5.C_6H_{13}$), having equal numbers of aliphatic carbon atoms (C_6H_{13}) and aromatic carbon atoms (C_6H_5), might have an equal chance of occurring in the *saturates fraction* or the *aromatics fraction*. It might also be argued that tridecylbenzene ($C_{13}H_{27}.C_6H_5$), having more aliphatic ($C_{13}H_{27}$) than aromatic (C_6H_5) features, has a greater likelihood of occurring in the *saturates fraction* than in the *aromatics fraction*. However, the partition of either of these hydrocarbons into the *saturates fraction* or the *aromatics fraction* depends upon the nature of the adsorbent and the mechanics of the separation procedure. Thus, the nomenclature of convenience is not necessarily correct in a chemical sense.

Another two operational definitions are those of the *carbenes* and the *carboids* (Chapter 1). Both are, by definition, insoluble in benzene (or toluene), but the carbenes are soluble in carbon disulfide (or pyridine) whereas the carboids are insoluble in carbon disulfide (or pyridine) (Speight, 1991). Only traces, if any, of the carbenes and carboids occur in natural petroleum and any quantities therein are not sufficient to alter the character of the petroleum or the character of the asphaltene fraction if the benzene treatment of the feedstock is omitted. Feedstocks that have received thermal treatment, such as occurs when *cracked residua* are produced, may contain measurable quantities of these materials. Carbenes and carboids were first named because of their relatively high carbon and low hydrogen content and are considered to be precursors of coke (Speight, 1987).

3.0 ASTM EVALUATION

In the early days of petroleum refining, it was considered sufficient to define feedstocks in terms of simple physical properties (Speight, 1981; Speight, 1991). The data obtained by physical methods were considered adequate for the prediction of feedstock behavior in refining operations. However, with the influx of different types of petroleum into refineries and the subsequent wide variations in behavior, there has been a need to define feedstocks in more detail in order to achieve some degree of predictability of feedstock behavior and product slate. This is especially true for the feedstocks that are the starting materials for a wide range of catalytic processes.

Catalyst deterioration during use, because of coke lay-down and metals deposition, is a continuing issue in catalyst-based processes. Frequent replacement of an expensive catalyst causes a major imbalance in process economics. Thus, a detailed knowledge of the nature of the feedstock constituents, through analytical and characterization procedures, is a necessary part of prerefining inspections.

In the early days of feedstock evaluation, many refiners developed their own methods of feedstock analysis and evaluation. Indeed, this trend has continued into modern refinery practice. Many of these methods have remained proprietary and, as a result, are not normally available. For example, instead of measuring the amount of asphaltenes in a feedstock, there may be reference to *modified naphtha insolubles* (*MNI*). This particular fraction might be equated to asphaltenes *if* the naphtha is equivalent to *n*-heptane, but usually it is not. A small percentage of aromatics and/or naphthenes and/or higher molecular weight paraffins can give the naphtha enhanced solvent properties, thereby changing the amount and character of the precipitate relative to *n*-heptane asphaltenes.

However, organizations such as the American Society for Testing and Materials (ASTM) in North America, the Institute of Petroleum (IP) in Britain, the Deutsche Industrie National in Germany, and the International Standards Organization in Switzerland have put considerable time and effort into the correlation and standardization of methods for the inspection, evaluation, and characterization of fossil fuels, particularly (in the present context) refinery feedstocks and refinery products.

As a result of these standard tests, a set of basic characteristics can be obtained which, if the procedure is followed rigorously, can be reproduced by different workers in various laboratories throughout the world. It is important that the results of any procedure be reproducible so that the separation becomes standardized and repeatable not only by different researchers but also in different laboratories. And many of the data can be correlated with feedstock suitability for various processing options as well as with product suitability for sale. This also requires comparison of the observed data with data obtained for other materials of known processability and character, because data obtained without any reference point are of only limited value. In keeping with their goals, the tests of all standards organizations require precision and reproducibility.

In this chapter, the various tests (with specific reference to the corresponding ASTM tests) that might be applied to the evaluation of feedstocks and products are referenced, along with other methods that are applied to feedstock/product characterization but are not yet recognized as standard tests. In fact, the latter tests often provide more detailed molecular information about the characteristics of feedstocks and products.

3.1 Ultimate Analysis

Chemically, petroleum is defined as a mixture of hydrocarbons containing small quantities of nitrogen-, oxygen-, and sulfur-containing compounds and traces of metallic constituents (Bestougeff, 1967; Colombo, 1967; Thornton, 1977). Compounds containing elements other than carbon and hydrogen are often referred to as heterocompounds or as heteroatom compounds. However, the presence of minor quantities of heterocompounds does not disqualify crude oil or natural gas from being a mixture of chemical hydrocarbons. Thus, petroleum has some value in the crude state and, when refined, provides fuel gas, petrochemical gas (methane, ethane, propane, and butane), transportation fuel (gasoline, diesel fuel, aviation fuel), solvents, lubricants, asphalt, and many other products.

The ultimate analysis (elemental composition) of petroleum is not reported to the same extent as for coal (Speight, 1991, 1994b). Nevertheless, there are ASTM procedures (ASTM, 1993) for the ultimate analysis of petroleum and petroleum products that allow comparisons of the various feedstocks and products (Speight, 1991). Many of the methods may have been designed for other materials but are applicable to petroleum with little or no modification.

For example, *carbon content* can be determined by the method designated for coal and coke (ASTM D-3178) or by the method designated for municipal solid waste (ASTM E-777). There are also methods designated for *hydrogen content* (ASTM D-1018, ASTM D-3178, ASTM D-3343, ASTM D-3701, and ASTM E-777), *nitrogen content* (ASTM D-3179, ASTM D-3228, ASTM D-3431, ASTM E-148, ASTM E-258, and ASTM E-778), *oxygen content* (ASTM E-385), and *sulfur content* (ASTM D-124, ASTM D-1266, ASTM D-1552, ASTM D-1757, ASTM D-2662, ASTM D-3177, ASTM D-4045, and ASTM D-4294).

In the data that are available (Speight, 1991), the proportions of the elements in petroleum vary only slightly over narrow limits (Chapter 1):

Carbon	83.0 – 87.0%
Hydrogen	10.0 – 14.0%
Nitrogen	0.1 – 2.0%
Oxygen	0.1 – 1.5%
Sulfur	0.1 – 6.0%

Yet, there is a wide variation in physical properties from the lighter, more mobile crude oils at one extreme to the heavier asphaltic crude oils at the other extreme. The majority of the more aromatic species and the heteroatoms occur in the higher boiling fractions of feedstocks. The heavier feedstocks are relatively rich in these higher boiling fractions.

3.2 Volatility

In very general terms, the higher the proportion of lower boiling fractions that can be used for liquid fuels, the higher the value of petroleum. Thus, the attractiveness of petroleum to the refiner is due to the fact that it can be subdivided by distillation into a variety of *hydrocarbon fractions* of different *cut points* (Speight, 1991). In fact, this was, and still is, the method by which petroleum feedstocks are evaluated for the application of various refinery options to produce a specific product or product slate.

As an early part of characterization studies, a correlation was observed between the quality of petroleum products and their hydrogen content, because gasoline, kerosene, diesel fuel, and lubricating oil are made up of hydrocarbon constituents containing high proportions of hydrogen. Thus, it is not surprising that tests to determine the volatility of petroleum and petroleum products were among the first to be defined. Indeed, volatility is one of the major tests for petroleum products, and it is inevitable that all products will, at some stage of their history, be tested for volatility characteristics.

The *vaporizing tendencies* of liquid fuels are the basis for the general characterization of petroleum fuels such as liquefied petroleum gas, natural gasoline, motor and aviation gasoline, naphtha, kerosene, diesel fuel, fuel oil, and gas oil (ASTM D-2715).

The *flash point* is the temperature to which the product must be heated under specified conditions to give off sufficient vapor to form a mixture with air that can be ignited momentarily by a flame (ASTM D-56, ASTM D-92, ASTM D-93). The *fire point* is the temperature at which petroleum or a petroleum product must be heated to burn continuously when a mixture of the vapor and air is ignited by a flame (ASTM D-92). The *vapor pressure* of petroleum and its constituent fractions is the force that must be exerted on the liquid to prevent it from vaporizing further (ASTM D-323).

Distillation involves the general procedure of vaporizing a liquid either at *atmospheric pressure* (ASTM D-86, ASTM D-216, ASTM D-285, ASTM D-447 and ASTM D-2892) or at *reduced pressure* (ASTM D-1160; ASTM D-196), and the data are reported in terms of one or more of the following items: (1) initial boiling point; (2) distillation temperature; (3) end point; (4) dry point; (5) percent distillate; (6) percent residue; (7) percent total recovery, i.e., the sum of the liquids and residue; (8) percent distillation loss; and (9) percent evaporated at a specific thermometer reading or other distillation temperature. The outcome is the generation of distillation proflies by which one crude oil can be compared with another (Chapter 1) and an assessment made of the potential of the crude oil(s) to generate distillate fuels. One of the methods (ASTM D-285) has been discontinued (in 1987) but, as often occurs when discontinuation of a method is announced, many laboratories will continue to use the technique for some time to come as means of deriving distillation profiles.

Detailed fractionation of the sample might be of secondary importance. Thus, it must be recognized that the general shape of a one-plate distillation curve is often adequate for making engineering calculations, correlating with other physical properties, and predicting the product slate (Nelson, 1958).

Another method that is increasingly popular for application to a variety of feedstocks and is commonly known as *simulated distillation* (ASTM D-2887). The method has been well researched in terms of method development and application (Hickerson, 1975; Green, 1976; Stuckey, 1978; Vercier and Mouton, 1979; Thomas et al., 1983; Romanowski and Thomas, 1985; MacAllister and DeRuiter, 1987; Schwartz et al., 1987; Thomas et al., 1987).

The benefits of the technique include good comparisons with other ASTM distillation data as well as the application to higher boiling fractions of petroleum. In fact, the data output includes the provision of the corresponding Engler profile (ASTM D-86) as well as the prediction of other properties such as vapor pressure and flash point (DeBruine and Ellison, 1973). When it is necessary to monitor product properties, as is often the case during refining operations, such data provide a valuable aid to process control and on-line product testing.

For a more detailed analysis of feedstocks and products by distillation, a low-resolution, temperature-programmed gas chromatographic analysis has been developed to simulate the time-consuming true boiling point distillation. The method relies on the general observation that hydrocarbons are eluted from a nonpolar adsorbent in the order of their boiling points. The regularity of the elution order of the hydrocarbon components allows the retention times to be equated to distillation temperatures (Green et al., 1964), and the term *simulated distillation by gas chromatography* (or *simdis*) is used throughout the industry to refer to this technique.

Simulated distillation by gas chromatography is often applied in the petroleum industry to obtain true boiling point data for distillates and crude oils (Butler, 1979). Two standardized methods (ASTM D-2887 and ASTM D-3710) are available for determining the boiling point of petroleum fractions and gasoline, respectively. The ASTM D-2887 method utilizes nonpolar, packed gas chromatographic columns in conjunction with flame ionization detection. The upper limit of the boiling range covered by this method is approximately 540°C (1000°F) atmospheric equivalent boiling point. Efforts in which high-temperature gas chromatography were used have focused on extending the scope of the ASTM D-2887 method for higher boiling petroleum materials to 800°C (1470°F) atmospheric equivalent boiling point (Schwartz et al., 1988).

3.3 Density and Specific Gravity

In the early years of the petroleum industry, density and specific gravity (ASTM D-287, ASTM D-1298, ASTM D-941, ASTM D-1217, and ASTM D-1555) were the principal analytical specifications for feedstocks and refinery products. They were used to give an estimate of the amount of kerosene in crude oil. More complex relationships also provided a guide to the proportions of the other constituents in petroleum (Speight, 1991).

Specific gravity, which approximates closely (but not exactly) to *density*, is the most common term at present but suffers from the disadvantage that a very narrow range of values applies to a wide range of feedstocks (see, for example, Speight, 1991). However, the American Petroleum Institute devised a wider scale that is based on the inverse of the specific gravity scale (the *API gravity*):

$$^{\circ}\text{API} = (141.5/\text{sp. gr. @ } 60/60\text{ F}) - 131.5$$

It is possible to recognize certain trends between the API gravity and other properties, such as sulfur content, nitrogen content, viscosity, carbon residue, and asphaltene or asphaltic (asphaltene plus resin) content (Speight, 1981; 1991).

3.4 Viscosity

Viscosity is a measure of the ability of a liquid to flow or a measure of its resistance to flow. The specification for petroleum products may include a viscosity term or relationship (e.g., lubricating oil), and there is considerable variation in the viscosities of individual petroleum fractions (Speight, 1991) and the different types of petroleum, such as a bitumen (Athabasca), two heavy oils (Cold Lake and Lloydminster), and conventional (light) crude oil (Figure 2-1). In addition, viscosity data are often employed to indicate what thermal forces will be necessary to move petroleum products and crude petroleum around a refinery.

There are several methods for the determination of *viscosity,* and the units of viscosity vary depending upon the method (ASTM D-445, ASTM D-88, ASTM D-2161, ASTM D-341, and ASTM D-2270). Viscosity data may also be reported as Saybolt universal seconds (SUS) or as Saybolt furol seconds (SFS), which, by means of conversion factors, can be reported as stokes; it is possible to convert the several viscosity scales (Speight, 1991), especially Saybolt viscosity to kinematic viscosity (ASTM D-2161).

Viscosity is often expressed in units of *stokes* or in units of *poises* (*poises* = *stokes* x *specific gravity*), and the viscosity of crude oils may vary from several stokes for liquid products and the lighter crude oils to several thousand stokes for heavier crude oils and even to several million stokes for residua and bitumen.

3.5 Carbon Residue

All thermal processes wherein the higher molecular weight constituents of petroleum, especially the higher molecular weight constituents of heavy oils and residua, are decomposed in the presence or absence of catalyst and/or hydrogen, produce coke in addition to the desired lower boiling liquid products (Nelson, 1976; Speight, 1991). Therefore, tests have been designed to determine the amount of *thermal coke* (*carbon residue*) that is likely to be produced by a feedstock (Nelson, 1964; Nelson, 1978). However, it must be remembered that

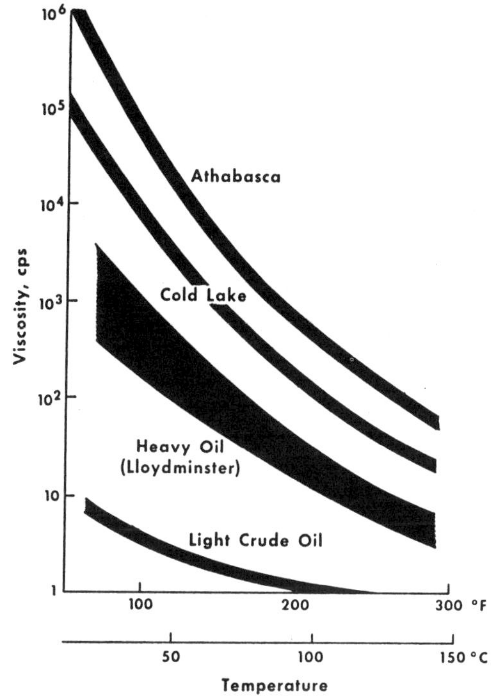

Figure 2-1. Variation in the viscosities of a bitumen (Athabasca), two heavy oils (Cold Lake and Lloydminster), and conventional (light) crude oil with temperature.

the amount of coke produced may be an indication only of the amount of coke produced in a thermal process. In fact, the carbon residue is a measure of the *coke-forming propensity* of the material under investigation.

There are two older well-used methods for determining the carbon residue: the *Conradson method* gives the Conradson carbon residue (CCR) (ASTM D-189) and the *Ramsbottom method* gives the Ramsbottom carbon residue (RCR) (ASTM D-524). The difference between the two methods is reflected in the manner in which the sample is heated to the final temperature. The methods are equally applicable, and it is possible to interconnect the data (Speight, 1991). Recently, a newer method has also been accepted (ASTM D-4530) that requires smaller sample amounts and was originally developed as a *thermogravimetric method.* The carbon residue produced by this method is often referred to as the microcarbon residue (MCR). Agreements between the data from the three methods are good, making it possible to interrelate all of the data from carbon residue tests (Long and Speight, 1989).

Determination of the carbon residue of a lower boiling fraction of petroleum, or of lighter petroleum, may be difficult because of the high volatility of the material and the low yield of the carbon residue. To overcome the extremely small values of carbon residue when the tests are applied to such materials, it is customary to distill such products to a 10% residue and determine the carbon residue thereof.

There are general relationships (Nelson, 1974; Speight, 1991) between the carbon residue of a feedstock and parameters such as the asphaltene content and the API gravity, which extends the relationship of the carbon residue to several other feedstock parameters such as sulfur content, nitrogen content, metals content, and viscosity (Speight, 1981; Speight, 1991). There may even be a more precise relationship between carbon residue and hydrogen content, H/C atomic ratio, nitrogen content, and sulfur content that can be used to provide more definitive information about the anticipated behavior of a variety of different feedstocks in thermal processes (Roberts, 1989).

3.6 Metals Content

Petroleum generally contains *metallic constituents* (Chapter 3), and this is particularly so for the heavier feedstocks and nonvolatile residua. These constituents, of which nickel and vanadium are the principal metals, strongly influence feedstock behavior in refining. Deposition of these metals in any form on the catalyst leads to catalyst deactivation, whether by physical blockage of the pores or by destruction of reactive sites. The metallic constituents may also act as nuclei and promote coke formation during thermal processes.

In the present context, the metals must first be removed if erroneously high carbon residue data are to be avoided. Alternatively, they can be estimated as ash by complete burning of the coke after carbon residue determination.

Many methods have been proposed by the ASTM for metals detection in petroleum and petroleum products. One particular method (ASTM D-3605) is the most general of these methods, and the latest ASTM volumes should be consulted for more detail.

3.7 Separation Methods

In very general terms, petroleum is a mélange of (1) hydrocarbons, (2) nitrogen compounds, (3) oxygen compounds, (4) sulfur compounds, and (5) metallic constituents (Chapter 1). However, this general definition is not adequate to describe the true composition as it relates to the behavior of petroleum during refining. For example, the occurrence of amphoteric species (i.e., compounds having a mixed acid-base nature) is not always addressed, nor is the phenomenon of molecular size or the occurrence of specific functional types that can play a major (often adverse) role in refining processes.

The amounts of hydrogen needed to upgrade feedstocks to the various products can, indeed, be estimated from the H/C atomic ratio, e.g.,

$$\underset{\text{feedstock}}{C_nH_{2n-8}} + 6H_2 \rightarrow \underset{\text{product}}{C_nH_{2n+2}}$$

but the lack of knowledge about what might, and can, happen due to the complexity of the chemical reactions leaves the estimates subject to grievous error. Much more information is needed to understand the processability, specifically the hydrogen requirements, of the feedstock. However, one must be careful that the characterization studies are not allowed to exert too much control over process definition; they are to be used in a supporting role!

Such information can be provided by analytical fractionation of feedstocks (Chapter 1) in order to study composition as it relates to refinery behavior and how one feedstock might vary from another (Speight, 1991). One apparent drawback to this work is the use of a variety of methods, which makes interlaboratory and even intralaboratory comparisons difficult unless standardized procedures are employed.

It is also essential, in work of this type, to determine the goals of the separation and then to choose an appropriate method by which the goals can be achieved. The selection of any separation procedure depends primarily on what information about the feedstock is desired. For example, separation into multiple fractions to examine the minute details of feedstock composition will require a more complex sequence of steps than the separation of only four bulk fractions.

Feedstock and product evaluation by separation into various fractions has been used quite successfully for several decades. Knowledge of the components of a feedstock on a *before-refining* and *after-refining* basis has been a valuable aid to process development (Rostler, 1965; Riediger, 1971; Altgelt and Gouw, 1979; Gary and Handwerk, 1984; Speight, 1991).

There are several ASTM procedures for feedstock and product evaluation, such as (1) the determination of aromatic content of olefin-free gasoline by silica gel adsorption (ASTM D-936), (2) the separation of aromatic and non-aromatic fractions from high-boiling oils (ASTM D-2549), (3) the determination of hydrocarbon groups in rubber extender oils by clay-gel adsorption (ASTM D-2007), and (4) the determination of hydrocarbon types in liquid petroleum products by a fluorescent indicator adsorption test (ASTM D-1319).

3.8 Other Tests

Several other tests are often employed as a guide to feedstock/product behavior and find use in the evaluation of feedstocks and products. These will not be described in detail here but are included by name and number for reference purposes.

These tests are: (1) *aniline point* (ASTM D-611 and ASTM D-801), (2) *liquefaction* and *solidification* (ASTM D-87, ASTM D-127, ASTM D-2500, and ASTM D-3117), (3) *pour point* (ASTM D-97), (4) *dropping point* (ASTM D-566 and ASTM D-2265), (5) *congealing point* (ASTM D-938), (6) *surface tension* and *interfacial tension* (ASTM D-971), (7) *specific heat* (ASTM D-2766), (8) *electrical conductivity* (ASTM D-3114), and (9) *refractive index* (ASTM D-1218 and ASTM D-2159).

It is appropriate here to mention the use of gas chromatography as an analysis technique for identifying the individual constituents of feedstocks and products. The technique has proved to be an exceptional and versatile instrumental tool for analyzing volatile products. However, use of this technique for direct feedstock analysis, especially analysis of the higher molecular weight fractions of petroleum, is not only difficult but also subject to many limitations in addition to the possibility of thermal decomposition before volatilization occurs.

A more recent and very important development in gas chromatography is its combination with a mass spectrometer as the detector. The technique in which gas chromatography is combined with mass spectrometry (GC/MS) has proved to be a powerful tool for identifying many compounds at very low levels in a wide-range-boiling matrix (Poirier and Das, 1984; Lin et al., 1989). By the combination of the two techniques in one instrument, the onerous trapping of fractions from the gas chromatographic column is avoided, and higher sensitivities can be attained. Because of the much reduced complexity of the gas chromatographic fractions, mass spectrometer scans carried out at regular intervals yield simpler spectra, from which compound classes can now be determined more easily.

Pyrolysis gas chromatography can be used to obtain information on the gross composition of heavy petroleum fractions. In this technique, the sample under investigation is pyrolyzed and the products are introduced into a gas chromatography system for analysis. The technique also has the potential to be employed as an indicator of the yields and efficiency of refining processes.

4.0 SPECTROSCOPIC METHODS

Spectroscopic studies have played an important role in the evaluation of petroleum and of petroleum products for the past three decades, and many of the methods are now used as standard methods of analysis for refinery feedstocks and products. Application of these methods to feedstocks and products is a natural consequence for the refiner.

The methods include the use of *mass spectrometry* to determine the (1) hydrocarbon types in middle distillates (ASTM D-2425), (2) hydrocarbon types of gas oil saturate fractions (ASTM D-2786), and (3) hydrocarbon types in low-olefin gasoline (ASTM D-2789), and (4) aromatic types in aromatic gas oil fractions (ASTM D-3239). *Nuclear magnetic resonance spectroscopy* has been developed as a standard method for the determination of hydrogen types in aviation turbine fuels (ASTM D-3701), and *x-ray fluorescence spectrometry* has been applied to the determination of lead in gasoline (ASTM D-2599) as well as of sulfur in various petroleum products (ASTM D-2622 and ASTM D-4294).

Infrared spectroscopy is used for the determination of benzene in motor and/or aviation gasoline (ASTM D-4053), and ultraviolet spectroscopy is used to evaluate mineral oils (ASTM D-2269) and to determine the naphthalene content of aviation turbine fuels (ASTM D-1840).

Other techniques include *flame emission spectroscopy* for determining trace metals in gas turbine fuels (ASTM D-3605); *absorption spectrophotometry* for the determination of the alkyl nitrate content of diesel fuels (ASTM D-4046); and *atomic absorption* as a means of measuring the lead content of gasoline (ASTM D-3237) and for the manganese content of gasoline (ASTM D-3831) as well as for determining the barium, calcium, magnesium, and zinc contents of lubricating oils (ASTM D-4628). Finally, *flame photometry* has been employed to measure the lithium/sodium content of lubricating greases (ASTM D-3340) and the sodium content of residual fuel oil (ASTM D-1318).

Nowhere is the contribution of spectroscopic studies more evident than in the delineation of structural types in the heavier feedstocks. This is important because the nature of these feedstocks has been unknown to refiners. One example is the *n.d.M. method* (ASTM D-3238), which is designed for the carbon distribution and structural group analysis of petroleum oils. Later investigators have taken structural group analysis several steps further than the n.d.M. method (Speight, 1991).

It is also appropriate at this point to give a brief description of other methods that are used for the identification of the constituents of petroleum (Yen, 1984). It is not the intent to indicate that any one of these methods can be used for identification purposes. However, although these methods may fall short of complete acceptability for the characterization of individual constituents of feedstocks, they can be used to obtain an overall evaluation of the feedstock in terms of molecular types.

4.1 Infrared Spectroscopy

Conventional infrared spectroscopy yields information about the functional features of various petroleum constituents. For example, infrared spectroscopy aids in the identification of N–H and O–H functions, the nature of polymethylene chains, the' C–H out-of-place bending frequencies, and the nature of any polynuclear aromatic systems (Yen, 1973).

With the recent progress in Fourier transform infrared (FTIR) spectroscopy, quantitative estimates of the various functional groups can also be made. This is particularly important for application to the higher molecular weight solid constituents of petroleum (i.e., the asphaltene fraction).

It is also possible to derive structural parameters from infrared spectroscopic data: (1) saturated hydrogen to saturated carbon ratio; (2) paraffinicity; (3) naphthenicity; (4) methyl group content; and (5) paraffin chain length. In conjunction with proton magnetic resonance (see next section), structural parameters such as the ratio of paraffinic methyl groups to aromatic methyl groups can be obtained.

4.2 Nuclear Magnetic Resonance

Nuclear magnetic resonance has frequently been employed for general studies and for structural studies of petroleum constituents (Bouquet and Bailleul, 1982; Hasan et al., 1989). In fact, proton magnetic resonance (PMR) studies (along with infrared spectroscopic studies) were perhaps the first studies of the modern era that allowed structural inferences about the polynuclear aromatic systems that occur in the high molecular weight constituents of petroleum.

In general, the proton (hydrogen) distribution in petroleum fractions can be subdivided into three types (Brown and Ladner, 1960) or into five types (Yen and Erdman, 1962).

The Brown and Ladner approach classifies the hydrogen types into (1) aromatic ring hydrogen, (2) aliphatic hydrogen adjacent to an aromatic ring, and (3) aliphatic hydrogen remote from an aromatic ring. The Yen and Erdman approach subdivides the hydrogen distribution into (1) aromatic hydrogen, (2) substituted hydrogen next to an aromatic ring, (3) naphthenic hydrogen, (4) methylenic hydrogen, and (5) terminal methyl hydrogen remote from an aromatic ring. Other ratios are also derived from which a series of structural parameters can be calculated.

However, it must be remembered that the structural details of the carbon backbone obtained from proton spectra are derived by inference and that protons at peripheral positions can be obscured by intermolecular interactions. This, of course, can cause errors in the ratios that can have a substantial influence on the outcome of the calculations (Ebert et al., 1987; Ebert, 1990).

It is in this regard that carbon-13 magnetic resonance (CMR) can play a useful role. Since carbon magnetic resonance is used to analyze the carbon distribution types, the obvious structural parameter to be determined is the aromaticity, f_a. Direct determination from the various carbon type environments is one of the better methods for determining aromaticity (Snape et al., 1979). Thus, through a combination of proton and carbon magnetic resonance techniques, the structural parameters can be refined and, by using the solid-state high-resolution CMR technique, additional structural parameters can be obtained (Weinberg at al., 1981).

4.3 Mass Spectrometry

Mass spectrometry can play a key role in the identification of the constituents of feedstocks and products (Aczel, 1989). The principal advantages of mass spectrometric methods are (1) high reproducibility of quantitative analyses, (2) the potential for obtaining detailed data on the individual components and/or carbon number homologs in complex mixtures, and (3) the requirement for a minimal sample size for analysis. The ability of mass spectrometry to identify individual components in complex mixtures is unmatched by that of any other modern analytical technique. Perhaps the exception is gas chromatography.

However, there are disadvantages arising from the use of mass spectrometry: (1) the limitation of the method to organic materials that are volatile and stable at temperatures up to 300°C (570°F) and (2) the difficulty in separating isomers for absolute identification. The sample is usually destroyed, but this is seldom a disadvantage.

In spite of these limitations, mass spectrometry does furnish useful information about the composition of feedstocks and products, even if this information is not as exhaustive as might be desired. There are structural similarities that might hinder identification of individual components and, consequently, make identification by type or by homolog more meaningful, since similar structural types may be presumed to behave similarly in processing situations and knowledge of the individual isomeric distribution may add only a little to an understanding of the relationships between composition and processing parameters.

Mass spectrometry should be used discriminately where a maximum amount of information can be expected. The heavier nonvolatile feedstocks are, for practical purposes, beyond the useful range of routine mass spectrometry. At the elevated temperatures necessary to encourage volatility, thermal decomposition would occur in the inlet and any subsequent analysis would be biased to the low molecular weight end and to the lower molecular products produced by the thermal decomposition.

5.0 MOLECULAR WEIGHT

Even though refining produces, in general, species of lower molecular weight than those originally in the feedstock, there is still a need to determine the molecular weights of the original constituents as well as those of the products as a means of understanding the process. For the original constituents and products, e.g., resins and asphaltenes, that have little or no volatility, *vapor pressure osmometry* (VPO) has been proved to be quite useful.

A particularly appropriate adaptation of the method involves the use of at least two different solvents, and the data are then extrapolated to infinite dilution (Schwager et al., 1979). Different temperatures have also been used for a particular solvent, after which the data are extrapolated to room temperature (Moschopedis and Speight, 1977; Speight et al., 1985; Speight, 1991).

In this manner, different solvents are employed and the molecular weight of a petroleum fraction or product can be determined, for which it can be assumed that there is little, or no, influence from any intermolecular forces. In fact, it is strongly recommended that to negate concentration effects and temperature effects the molecular weight determination be carried out at three different concentrations at three different temperatures. The data for each temperature are then extrapolated to zero concentration, and the zero concentration data at each temperature are then extrapolated to room temperature (Speight, 1987; Speight, 1991).

6.0 USE OF ANALYTICAL DATA

The data derived from the analytical procedures can be employed to give the refiner an indication of the behavior of the feedstock during refining (Gary and Handwerk, 1984) and to predict process options that may lead to improvements in product character and/or yield. With the complexity of modern refinery feedstocks, there are disadvantages in relying on the use of bulk properties as the sole means of predicting current-day feedstock behavior (Dolbear et al., 1987). Other means of evaluation are necessary, and this is where the fractionation data can be a very valuable aid to understanding feedstock behavior and process options.

A correlation between the hyrogen/carbon atomic ratio and the microcarbon residue illustrates the tendency for microcarbon to increase with an increase in aromaticity of the fraction (Figure 2-2). It must be remembered that as the aromaticity of a petroleum fraction increases, there is a corresponding increase in the size of the polynuclear aromatic system and in the heteroatom content. Both are contributors to nonvolatility and the propensity to form thermal coke. A more revealing correlation is that between the microcarbon residue and the molecular weight (Figure 2-3). This correlation reveals that association of the asphaltenes, if it occurs at all in the crude oil, does not appear to influence the propensity for coke formation at the temperatures (ca. 500°C) of the microcarbon test and, by inference, at process temperatures.

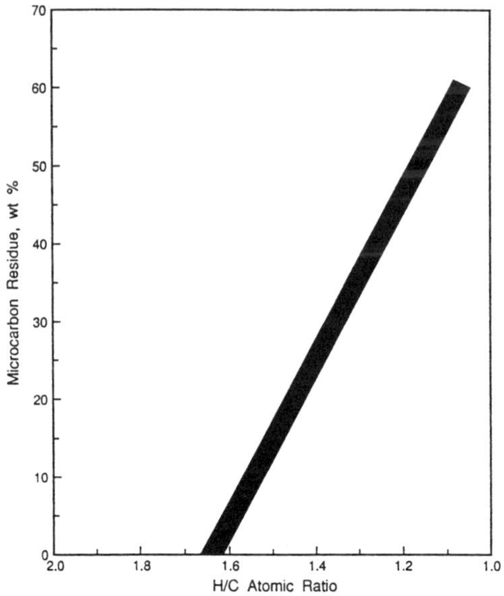

Figure 2-2. Relationship of microcarbon residue to aromaticity of petroleum fractions.

By careful selection of an appropriate technique, it is possible to obtain a detailed overview of feedstock and product composition that can be used for process predictions. Using the adsorbent separation as an example (Corbett and Petrossi, 1978), it becomes possible to develop a *petroleum map* (Long and Speight, 1989).

It must be recognized that such a map does not give any indication of the complex interactions that occur between, for example, fractions such as the asphaltenes and resins (Koots and Speight, 1975), but it does allow predictions of feedstock behavior. It must also be recognized that such a representation will vary for different feedstocks.

To proceed from bulk property data to full-scale production is not the preferred step. Further evaluation of the processability of the feedstock is usually necessary through the use of feedstock/product composition data that will, in conjunction with bulk property data, be of substantial aid in moving to pilot scale operations. It may then be possible to develop correlations between the data obtained from the actual plant operations (as well as the pilot plant data) and one or more of the physical properties determined as part of the initial evaluation.

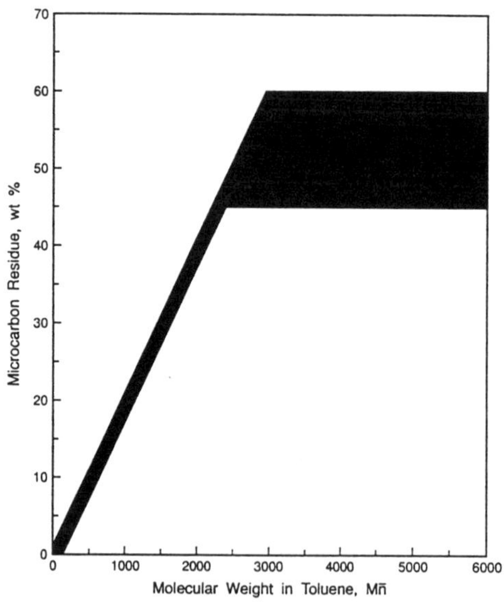

Figure 2-3. Relationship of microcarbon residue to molecular weight of petroleum fractions.

In summary, the use of analytical data has been, and remains, an integral part of refinery operations. There is a need to understand the nature of the constituents of the heavier feedstocks in more detail in order to be able to predict product yield and product distribution. The analytical methods currently at hand are an aid to accomplishing this goal. But it is only when a full understanding of the character of the heavier feedstocks is available that confidence in prediction will emerge.

The use of analytical data does not stop with the definition or evaluation of the character of the feedstock. The process continues after refining and at the time when products are being prepared for sale.

Most petroleum products must meet specifications before sale can occur. This is very true for the liquid fuels, asphalt, and coke, which have been subject to stringent testing for decades (Guthrie, 1960; Batts and Fathoni, 1991; Mushrush and Speight, 1995). The performance of fuels in service, asphalt in road use, and coke during combustion (with particular emphasis on the sulfur content and the generation of sulfur oxides) has been the subject of many studies.

The tests are used to develop an understanding of the principal factors that determine the behavior of petroleum and petroleum products during refining and use.

7.0 REFERENCES

Aczel, T. 1989. Preprints Div. Petrol. Chem. Am. Chem. Soc. 34(2): 318.

Alfredson, T. V. 1981. J. Chromatogr. 218: 715.

Altgelt, K. H. 1968. Preprints Div. Petrol. Chem. Am. Chem. Soc. 13(3): 37

Altgelt, K. H. 1970. Bitumen, Teere, Asphalte, Peche. 21: 475.

Altgelt, K. H., and Gouw, T. H. 1975. In Advances in Chromatography. J.C. Giddings, E. Grushka, R.A. Keller, and J. Cazes (Editors). Marcel Dekker Inc., New York.

Altgelt, K. H., and Gouw, T. H. 1979. Chromatography in Petroleum Analysis. Marcel Dekker Inc., New York.

ASTM. 1993. Annual Book of Standards. American Society for Testing and Materials, Philadelphia, PA.

Baltus, R. E., and Anderson, J. L. 1984. Fuel. 63: 530.

Batts, B. D., and Fathoni, A. Z. 1991. Energy & Fuels. 5: 2.

Bestougeff, M. 1967. In Fundamental Aspects of Petroleum Geochemistry. B. Nagy and U. Colombo (Editors). Elsevier, Amsterdam. Chapter 3.

Bollet, C., Escalier, J. C., Souteyrand, C., Caude, M., and Rosset, R. 1981. J. Chromatogr. 206: 289.

Bouquet, M., and Bailleul, A. 1982. In Petroananlysis '81. Advances in Analytical Chemistry in the Petroleum Industry 1975–1982. G. B. Crump (Editor). John Wiley & Sons, Chichester, England.

Boyd, M. L., and Montgomery, D. S. 1963. J. Inst. Petrol. 49: 345.

Brown, J. K., and Ladner, W. R. 1960. Fuel. 39: 87.

Butler, R. D. 1979. In Chromatography in Petroleum Analysis. K. H. Altgelt and T. H. Gouw (Editors). Marcel Dekker Inc., New York.

Chartier, P., Gareil, P. Caude, M., Rosset, R., Neff, B., Bourgognon, H. F., and Husson, J. F. 1986. J. Chromatogr. 357: 381.

Chmielowiec, J., Beshai, J. E., and George, A. E. 1980. Fuel. 59: 838.

Colin, J .M., and Vion, G. 1983. J. Chromatogr. 280: 152.

Colombo, U. 1967. In Fundamental Aspects of Petroleum Geochemistry. B. Nagy and U. Colombo (Editors). Elsevier, Amsterdam. Chapter 8.

Corbett, L. W., and Petrossi, U. 1978. Ind. Eng. Chem. Prod. Res. Dev. 17: 342.

Coulombe, S., and Sawatzky, H. 1986. Fuel. 65: 552.

DeBruine, W., and Ellison, R. J. 1973. J. Petrol. Inst. 59: 146.

Dolbear, G. E., Tang, A., and Moorehead, E. L. 1987. In Metal Complexes in Fossil Fuels. R. H. Filby and J. F. Branthaver (Editors). Symposium Series No. 344. American Chemical Society, Washington, DC. p. 220.

Drushel, H. V. 1983. J. Chromatogr. Sci. 21: 375.

Drushel, H.V., and Sommers, A.L. 1966. Anal. Chem. 38: 19.

Ebert, L.B. 1990. Fuel Sci. Technol. Int. 8: 563.

Ebert, L. B., Mills, D. R., and Scanlon, J. C. 1987. Preprints Div. Petrol. Chem. Am. Chem. Soc. 32(2): 419.

Felix, G., Bertrand, C., and Van Gastel, F. 1985. Chromatographia. 20(3): 155.

Francisco, M. A., and Speight, J. G. 1984. Preprints Am. Chem. Soc. Div. Fuel Chem. 29(1): 36.

Gary, J. H., and Handwerk, G. E. 1984. Petroleum Refining: Technology and Economics. Second Edition. Marcel Dekker Inc., New York.

George, A. E., and Beshai, J. E. 1983. Fuel. 62: 345.

Green, L. E. 1976. Hydrocarbon Process. 55(5): 205.

Green, L. E., Schmauch, L. J., and Worman, J. C. 1964. Anal. Chem. 36: 1512.

Guthrie, V. B. (Editor). 1960. Petroleum Products Handbook. McGraw-Hill, New York.

Hasan, M., Ali, M. F., and Arab, M. 1989. Fuel. 68: 801.

Hausler, D. W., and Carlson, R. S. 1985. Preprints Div. Petrol. Chem. Am. Chem. Soc. 30(1): 28.

Hayes, P. C., and Anderson, S. D. 1987. J. Chromatogr. 387: 333.

Hickerson, J. F. 1975. In Special Publication No. STP 577. American Society for Testing and Materials, Philadelphia. p. 71.
Jewell, D. M., Albaugh, E. W., Davis, B. E., and Ruberto, R. G. 1974. Ind. Eng. Chem. Fundam. 13: 278.
Koots, J. A., and Speight, J. G. 1975. Fuel. 54: 179.
Lin, D. P., Litorja, L. A., and Abbas, N. M. 1989. Fuel. 68: 257.
Long, R. B., and Speight, J. G. 1989. Rev. Inst. Fr. Petrole. 44: 205.
Lundanes, E., and Greibrokk, T. 1985. J. Chromatogr. 349: 439.
Lundanes, E., Iversen, B., and Greibokk, T. 1986. J. Chromatogr. 366: 391.
MacAllister, D. J., and DeRuiter, R. A. 1985. Paper SPE 14335. 60th Annual Technical Conference. Society of Petroleum Engineers. Las Vegas. September 22–25.
Matsushita, S., Tada, Y., and Ikushige. 1981. J. Chromatogr. 208: 429.
McKay, J. F., Cogswell, T. E., and Latham, D. R. 1974. Preprints Div. Petrol. Chem. Am. Chem. Soc. 19(1): 25.
McKay, J. F., Amend, P. J., Cogswell, T. E., Harnsberger, P. M., Erickson, R. B., and Latham, D. R. 1977. Preprints Am. Chem. Soc. Div. Petrol. Chem. 22(2): 708.
Miller, R. L., Ettre, L. S., and Johansen, N. G. 1983. J. Chromatogr. 259: 393
Moschopedis, S. E., and Speight, J. G. 1977. Fuel. 56: 344.
Mushrush, G. W., and Speight, J. G. 1995. Petroleum Products: Instability and Incompatibility. Taylor & Francis, Washington, DC.
Nelson, W. L. 1958. Petroleum Refinery Engineering. McGraw-Hill. New York.
Nelson, W. L. 1964. Oil and Gas J. 62(6): 124.
Nelson, W. L. 1974. Oil and Gas J. 72(6): 72.
Nelson, W. L. 1976. Oil and Gas J. 74(21): 60.
Nelson, W. L. 1978. Oil and Gas J. 76(41): 71.
Norris, T. A., and Rawdon, M. G. 1984. Anal. Chem. 56: 1767.
Oelert, H. H. 1969. Erdoel und Kohle. 22: 536.
Oudin, J. L. 1970. Rev. Inst. Fr. Petrole. 2: 470.
Poirier, M. A., and Das, B. S. 1984. Fuel. 63. 361.
Rawdon, M. 1984. Anal. Chem. 56: 831.
Reynolds, J. G., and Biggs, W. R. 1988. Fuel Sci. Technol. Int. 6: 329.
Riediger, B. 1971. Die Verarbeitung des Erdoles. Springer-Verlag, Berlin.
Roberts, I. 1989. Preprints Am. Chem. Soc. Div. Petrol. Chem. 34(2): 251.
Romanowski, L. J., and Thomas, K. P. 1985. Report No. DOE/FE/60177-2326. United States Department of Energy, Wahington, D.C.
Rostler, F. S. 1965. In Bituminous Materials: Asphalt, Tars and Pitches. A. Hoiberg (Editor). Interscience, New York, 11(1): 151.
Schwager, I., Lee, W. C., and Yen, T. F. 1979. Anal. Chem. 51: 1803.

Schwartz, H. E., and Brownlee, R. G. 1986. J. Chromatogr. 353: 77.
Schwartz, H. E., Brownlee, R. G., Boduszynski, M. M., and Su, F. 1987. Anal. Chem. 59: 1393.
Schwartz, H. E., Levy, J. M., and Guzowski, J. P. 1988. In Modern Supercritical Fluid Chromatography. C. M. White (Editor). Huthig Verlag, New York.
Skripek, M., Alley, S. K., Helfrey, P. F., and Leuth, P. F. 1975. Proc. 79th National Meeting, American Institute of Chemical Engineers.
Smith, R. D., Wright, B. W., and Yonker, C. R. 1988. Anal. Chem. 60: 1323.
Snape, C. E., Ladner, W. R., and Bartle, K. D. 1979. Anal. Chem. 51: 2189.
Snyder, L. R., and Buell, B. E. 1965. Anal. Chim. Acta. 33: 285.
Speight, J. G. 1979. Information Series No. 84. Alberta Research Council, Edmonton, Alberta, Canada.
Speight, J. G. 1981. The Desulfurization of Heavy Oils and Residua. Marcel Dekker Inc., New York.
Speight, J. G. 1986. Preprints Am. Chem. Soc. Div. Petrol. Chem. 31(4): 818.
Speight, J. G. 1987. Preprints Am. Chem. Soc. Div. Petrol. Chem. 32(2): 413.
Speight, J. G. 1991. The Chemistry and Technology of Petroleum. 2nd Edition. Marcel Dekker Inc., New York.
Speight, J. G. 1994a. Arabian J. Sci. Eng. 19(2B): 335.
Speight, J. G. 1994b. The Chemistry and Technology of Coal. 2nd Edition. Marcel Dekker Inc., New York.
Speight, J. G., Wermick, D. L., Gould, K. A., Overfield, R. E., Rao, B. M. L., and Savage, D. W. 1985. Rev. Inst. Fr. Petrole. 40: 27.
Stuckey, C. L. 1978. J. Chromatogr. Sci. 16:482.
Suatoni, J. C., and Garber, H. R. 1975. J. Chromatogr. Sci. 13: 367.
Swain, E. J. 1991. Oil and Gas J. 89(36): 59.
Thomas, K. P., Barbour, R. V., Branthaver, J. F., and Dorrence, S. M. 1983. Fuel. 62: 438.
Thomas, K. P., Harnsberger, P. M., and Guffey, F. D. 1987. Report No. DOE/MC/11076-2451. United States Department of Energy. Washington, DC.
Thornton, D. P. Jr. 1977. In Energy Technology Handbook. D. M. Considine (Editor). McGraw-Hill, New York. pp. 3–12.
Tissot, B. 1984. Characterization of Heavy Crude Oils and Petroleum Residues. Editions Technip, Paris.
Vercier, P., and Mouton, M. 1979. Oil and Gas J. 77(38): 121.
Weinberg, V. L., Yen , T. F., Gerstein, B. C., and Murphy, P. D. 1981. Preprints Div. Petrol. Chem. Am. Chem. Soc. 26(4): 816.
Yen, T. F. 1973. Fuel. 52: 93.
Yen, T. F. 1984. In The Future of Heavy Crude and Tar Sands. R. F. Meyer, J. C.

Wynn, and J. C. Olson (Editors). McGraw-Hill, New York.
Yen, T. F., and Erdman, J. G. 1962. Preprints Div. Petrol. Chem. Am. Chem. Soc. 7(3): 99.

Chapter 3

Metals and Heteroatoms in Heavy Crude Oils

John G. Reynolds
University of California
Lawrence Livermore National Laboratory
P. O. Box 808
Livermore
California 94551

1.0 INTRODUCTION

Heteroatoms (nitrogen, oxygen, sulfur, and metals) are found in every crude oil, and the concentrations have to be reduced to convert the oil to transportation fuel. The reason is that if nitrogen and sulfur are present in the final fuel during combustion, nitrogen oxides (NO_x) and sulfur oxides (SO_x) form, respectively. In addition, metals affect many upgrading processes adversely, poisoning catalysts in refining and causing deposits in combustion.

Although oxygen can have an important influence on petroleum properties, its removal during processing is somewhat less problematic than the removal of nitrogen and sulfur. Therefore, this chapter will focus on the effects of nitrogen, sulfur, and metals on petroleum properties and processing.

For heavy crude oil, the concentrations of sulfur are generally in the range 0.2 to 7.0 wt. %. Nitrogen concentrations vary from trace amounts to more than 1.0 wt. % and the concentrations of metals vary from a few parts per million to over 1000 ppm.

Typically heavy crude oils are distilled before use, which concentrates the heteroatoms into the nondistillable portion. It is important to note that the complexity of heavy crude oils increases with increasing boiling point and that the nondistillable portions are the most complex and most complicated to characterize (Chapter 1). In addition, the heteroatom constituents of crude oil may contain several heteroatoms per molecule.

Heteroatoms do affect every aspect of refining. Sulfur is usually the most concentrated and is fairly easy to remove; many commercial catalysts are available that routinely remove 90% of the sulfur. Nitrogen is more difficult to remove than sulfur, and there are fewer catalysts that are specific for nitrogen. Metals cause particular problems because they poison catalysts used for sulfur and nitrogen removal as well as other processes such as catalytic cracking.

There is a wealth of information about nitrogen, sulfur, vanadium, and nickel in heavy crude oils. However, the knowledge of the exact heteroatomic species in a fraction decreases as the boiling range of the fraction increases. This is primarily because the complexity of the fraction also increases. Knowing all the individual compounds that make up the heavy crude oil or a fraction is probably not possible at this time.

The following is a discussion of heteroatoms in heavy crude oil, focusing on their distributions in the whole oil and in the separated fractions, principally through their behavior during separation (such as distillation and chromatography), their molecular structure, and some molecular behavior during processing. This chapter is not intended to be a review or compendium of all the literature in the field, nor is inclusion or exclusion of any particular research an editorial comment. A large number of research reports were excluded because of logistics, and the data presented and discussed here were selected because they exemplify a particular point being presented.

2.0 HETEROATOM CONCENTRATIONS IN PETROLEUM

Sulfur and nitrogen concentrations are generally recognized to increase with increasing API gravity (Speight, 1991). There are some exceptions to this generalization, particularly when comparing crude oils that are derived from vastly different source rocks (Table 3-1). Crude oils from Saudi Arabia (Ali et al., 1985) show definite trends of increasing sulfur, nitrogen, nickel, and vanadium with decreasing gravity. Crude oils from China (Liang et al., 1991) also exhibit this trend for sulfur and vanadium, but not for nitrogen and nickel.

In addition, the absolute heteroatom concentrations for the crude oils from China do not compare with those for Arabian crude oils. Crude oils from California as well as from other parts of the world (Reynolds and Biggs, 1986; Reynolds, 1996; Dolbear et al., 1987; Chrones and Germain, 1989) further demonstrate that the heteroatom-gravity relationship is only cursory. However, the relationship is generally better when comparing crude oils produced within a given field from a specific source rock.

2.1 Heteroatom Distributions in Refinery Distillation Cuts

Kern River crude oil contains substantial amounts of heteroatoms in different fractions (Table 3-2) and, as anticipated, the sulfur concentration

Table 3-1. API gravities and heteroatom contents of selected crude oils (Ali et al., 1985; Chrones & Germain, 1989; Dolbear et al., 1987; Liang et al., 1991; Reynolds, 1996; Reynolds & Biggs, 1986).

Crude oil	API, °	S, wt %	N, wt %	Fe, ppm	Ni, ppm	V, ppm
Arabian E. Light	38.5	1.10	0.03	0	1	2
Arabian Light	33.8	1.81	0.07	1	3	16
Arabian Medium	30.4	2.59	0.09	1	8	28
Arabian Heavy	28.0	2.79	0.15	1	17	58
Shengli, China	13.8	0.82	0.72	na	51	3
Liaohe, China	17.9	0.26	0.41	na	38	1
Xinjiang, China	20.5	0.15	0.35	na	15	1
Henan, China	16.4	0.32	0.74	na	22	3
Hondo, California	19.2	4.93	0.62	0	68	220
Beta, California	16.2	3.58	0.81	43	84	135
Kern, California	13.6	1.10	0.7	28	63	35
Midway, California	11.8	1.54	0.63	na	53	64
Maya, Mexico	22.2	3.40	0.32	9	39	242
Athabasca, Canada	8	4.80	0.4	na	100	250
ANS, Alaska	26.2	1.1	0.2	1	11	23
Boscan, Venezuela	10.2	5.29	0.65	7	90	920

Table 3-2. Heteroatom contents of different fractions of Kern River crude oil; distillation by ASTM D-86 and D-1160.

Fraction	Crude	IBP to 400°F	400 to 500°F	500 to 650°F	650 to 800°F	800 to 1000°F	1000°F+
%	100	1.7	6.0	15.3	15.6	23.9	37.5
S, wt %	1.1	0.07	0.17	0.63	1.0	1.1	1.3
N, wt %	0.7	0.0	0.0	0.06	0.38	0.56	1.4
Fe, ppm	28	0	0	0	0	0	90
Ni, ppm	63	0	0	0	0	0	110
V, ppm	35	0	0	0	0	0	95

increases with the boiling range. Not much nitrogen is found in the fractions that boil below 260°C (500°F), and metals are generally found only in the nondistillable portion. Although Kern River has a lower sulfur content and higher nitrogen content than many heavy crude oils, the sulfur behavior during refinery-type distillation is representative of most heavy crude oils.

The lower boiling fractions of Arabian Heavy, Kern River, and Boscan crude oils account for little of the total sulfur; the nondistillable portion accounts for most of it (Figure 3-1). This is due to a combination of the higher sulfur

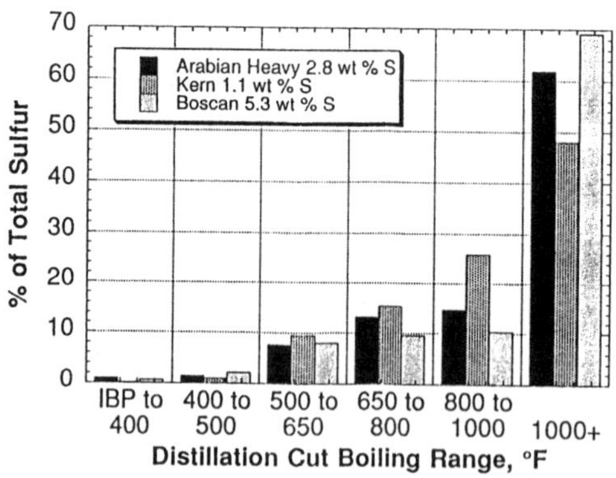

Figure 3-1. Distribution of sulfur in distillation fractions in selected crude oils.

concentrations of the higher boiling range fractions and the larger volume of these fractions in heavy crude oils.

A similar distribution is seen for nitrogen in Arabian Heavy, Kern River, and Boscan crude oils. Even lower concentrations of nitrogen are found in the light fractions of these crude oils, and the majority of the nitrogen is accounted for in fractions with boiling points above 345°C (650°F). In these three heavy crude oils, more than 70% of the nitrogen is found in the 540°C⁺ (1000°F) residuum. And, as illustrated for Kern River crude oil (Table 3-2), all of the metals are found in the 540°C (1000°F⁺) residuum of heavy crude oils. However, corrosion and entrainment sometimes cause metals to be found in fractions boiling below 540°C (1000°F).

2.2 Heteroatom Distribution in Molecular Distillation Fractions

As already noted, the majority of heteroatoms are found in the nondistillable portion or the 540°C⁺ (1000°F⁺) residuum. This fraction is nondistillable in the refinery and is above the practical range of regular atmospheric, vacuum, and analytical distillation because of the thermal decomposition that occurs in this temperature range. The invention of short-path or molecular distillation, however, has allowed deeper distillation cuts of this residual material to be produced (Vercier and Mounton, 1982).

Briefly, the unit uses a high vacuum (10^{-3} mm Hg or less) with a roller wiped-film system (which provides a thin layer of oil to be distilled) for heating the residuum. The condensing unit is within the mean-free-path length of the molecules (2 to 3 cm achieved with the high vacuum). This combination of thin film, short path, and heating with short residence time results in minimal decomposition of the crude oil.

Because the molecular distillation is a nonequilibrium process, the atmospheric equivalent boiling ranges of the fractions cannot be measured directly. The atmospheric equivalent boiling point (AEBP) concept was developed to compensate for this (Altgelt and Boduszynski, 1992, 1994) (Equations 1 and 2):

$$\text{mid-AEBP (°F)} = \{[(M_n - 140)(\text{sp. gr.})^{2.5}/3.40] \times 10^7\}^{1/3} \quad (1)$$

$$\text{mid-AEBP (°F)} = \{[(M_n - 170)(\text{H/C})^{0.9}/2.67] \times 10^7\}^{1/3} \quad (2)$$

The mid-AEBP is the temperature for the 50% mass point on the distillation curve of the fraction, and M_n is the molecular weight. Either the specific gravity (sp. gr.) or atomic hydrogen/carbon (H/C) ratio, which can be readily measured in the fractions can be used.

Using the mid-AEBP concept, even fractions of the nondistillable residue can be included in the boiling range curves and relationships such as the variation of sulfur and nitrogen contents can be derived as a function of the AEBP (Table 3-3). The molecular distillation is performed on the residuum portion of the Arabian Heavy because the high vacuum precludes distillation of the lighter ends. The material from the sequential elution fractionation (SEF) is due not to distillation but to solubilization of the nondistillable residue (boiling point >705°C^{+}; >1300°F^{+}). The fractions were produced by coating the 705°C^{+} (1300°F^{+}) residuum on a mixture of Teflon and glass beads and then passing various solvents of increasing solvent power (solubility parameter) over them to solubilize different classes of compounds. As the mid-AEBP increases, the sulfur and nitrogen concentrations of the fraction generally also increase. Also, the highest concentrations of both sulfur and nitrogen appear in the nondistillable fractions. This behavior follows the heteroatom behavior observed for refinery distillation cuts (see Table 3-2, for example). Thus, the higher the boiling range of the fraction, the higher the heteroatom concentration. This also establishes that the heteroatom concentration continues to increase as the volatility of the compounds decreases, which was not known previously for the 540°C^{+} (1000°F^{+}) residuum.

Table 3-3. Nitrogen and sulfur contents of Arabian Heavy 650°F^{+} residuum separated by molecular distillation and sequential elution fractionation as a function of atmospheric equivalent boiling point (Boduszynski, 1987).

Fraction	mid-AEBP[a], °F	wt % S	wt % N
Cut 1	662	2.62	0.03
Cut 2A	795	3.14	0.07
Cut 2B	890	3.36	0.10
Cut 3A	972	3.75	0.12
Cut 3B	1025	4.01	0.15
Cut 4A	1090	4.27	0.20
Cut 4B	1145	4.22	0.20
Cut 5A	1205	4.62	0.19
Cut 5B	1305	4.84	0.27
SEF-1	1560	5.45	0.39
SEF-2	2310	7.54	0.88
SEF-3	2570	8.02	0.98

a. Atmospheric equivalent boiling point.

In addition, this shows that the sulfur behavior (see Table 3-3) extends to other heavy crude oils, including Kern River, Arabian Heavy, Maya, Boscan, and Offshore California crude oils (Figure 3-2) (Altgelt and Boduszynski, 1994). Even though the concentrations of the heteroatoms in the crude oils are different, the behavior is basically the same. Kern River is a little different, as most fractions have similar sulfur concentrations. Note that the mid-AEBP concept extends the theoretical boiling range well above 1370°C (2500°F) and that the sulfur concentrations reflect the trends established by the volatile fractions. Nitrogen concentrations follow a similar trend (Altgelt and Boduszynski, 1994).

Metals are not found in any fractions that boil below 540°C (1000°F) (Figure 3-3) (Altgelt and Boduszynski, 1994). However, the molecular distillation allows volatilization of some of the metals in the fractions from 540 to 705°C (1000 to 1300°F). Some of these have been identified as the *etio* and *DPEP* homologous series of vanadyl petroporphyrins.

The sequential elution fractions show that the concentrations of vanadium in the nondistillable residues are very high and that the nickel concentration in these fractions shows the same behavior as vanadium: none in the <540°C

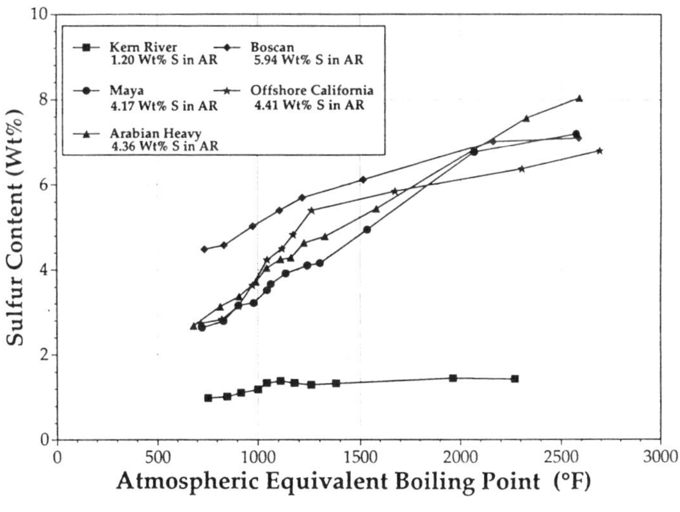

Figure 3-2. Sulfur concentrations in molecular distillation fractions and sequential elution fractions of selected residua (Altgelt and Boduszynski, 1994. Copyright Marcel Dekker, Inc. Reproduced with permission).

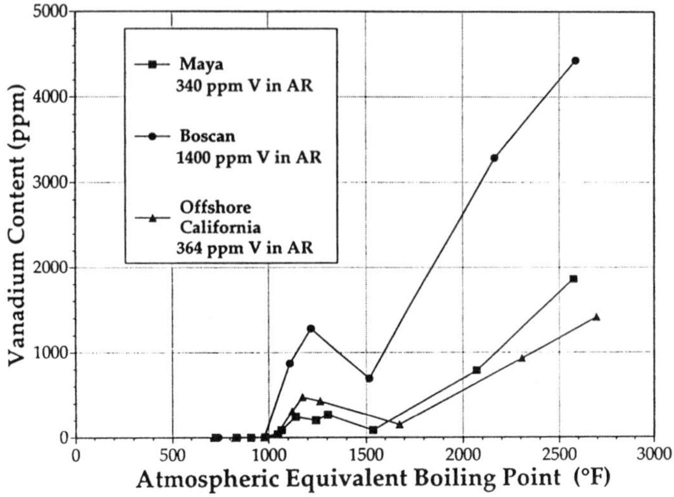

Figure 3-3. Vanadium concentrations in molecular distillation fractions and sequential elution fractions of selected residua (Altgelt and Boduszynski, 1994. Copyright Marcel Dekker, Inc. Reproduced with permission.)

(<1000°F) fraction, some in the 540 to 705°C (1000 to 1300°F) fraction, and high concentrations in the eluted SEF material. Although most crude oils have no iron, minor quantities of iron have been detected in the 540 to 705°C (1000 to 1300°F) fraction from Offshore California crude oil, but this iron is probably not bound in porphyrin systems.

This type of distillation and separation method, coupled with the AEBP concept, has led to a better understanding of the nature of heteroatoms in heavy crude oils, particularly in the residuum portion. Thus, the average number of nitrogen atoms per molecule in the molecular distillation and sequential elution fractions for Kern River, Maya, Arabian Heavy, Boscan, and Offshore California heavy crude oils can be estimated (Figure 3-4). As suspected from the heteroatom concentration data (Table 3-3 and Figures 3-2 and 3-3), the average number of nitrogen atoms per molecule increases with increasing mid-AEBP. More important, there can be as many as 10 nitrogen atoms in one molecule in the more polar sequential elution material. In addition, because this is an average parameter, the number of nitrogen atoms in a specific molecule could be a great deal higher.

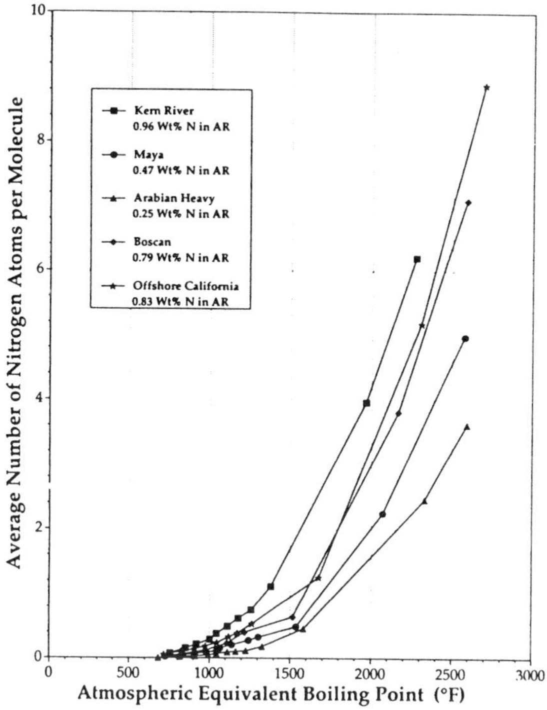

Figure 3-4. Average number of nitrogen atoms per molecule for selected residua (Altgelt and Boduszynski, 1994. Copyright American Chemical Society. Reproduced with permission).

The behavior of sulfur is similar to that of nitrogen, except for high-sulfur crude oils such as Boscan, Arabian Heavy, Maya, and Offshore California, in which the average number is 12 to 13 sulfur atoms per molecule.

2.3 Heteroatom Distributions in the SARA Separation

The saturates, aromatics, resins, and asphaltenes (SARA) separation method (Figure 3-5) (Chapter 1) was one of the first separation methods used for crude oils. As a result, there are many permutations in the literature, although it was adopted as a standard method of fractionation by the American Society for Testing and Materials (ASTM-D2007) (Speight, 1991). Even though the method is subject to criticism, this separation procedure essentially laid the framework for

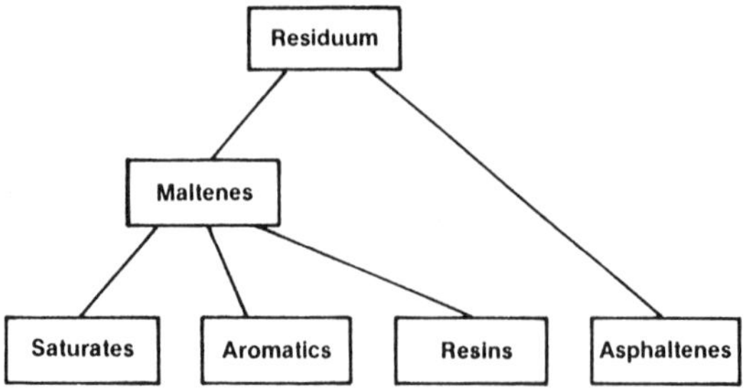

- **Asphaltenes**
  - **Hot Isooctane 40/1 by Vol Insoluble**
- **Maltenes**
  - **Hot Isooctane 40/1 by Vol Soluble**
- **Saturates**
  - **Elution Through Attapulgite Clay and Silica**
- **Aromatics**
  - **Elution Through Attapulgite Clay but Retained on Silica**
- **Resins**
  - **Retained on Attapulgite Clay**

Figure 3-5. Schematic representation of the saturates-aromatics-resins-asphaltenes separation (modified ASTM D-2007 procedure) (Reynolds, 1985. Copyright Marcel Dekker, Inc. Reproduced with permission).

more advanced and improved separation methods, so it will be mention briefly. In addition, asphaltene precipitation is commonly used in many characterization and processing studies (Chapters 4 and 5), so asphaltene properties also need some attention here.

The key to the separation is asphaltene precipitation, accomplished by adding large volumes of an alkane (Speight et al., 1984). This produces operationally defined fractions (Chapter 1): the insoluble portion, called the asphaltenes, and the soluble portion, called the maltenes (maltenes are further separated by column chromatography). The wide choice of precipitation solvents and column materials (Speight, 1991) indicates why the method has so many variations (Chapter 1).

After application of a separation method such as the SARA procedure, the polar fractions (the resins and asphaltenes) have the highest concentrations of sulfur and the saturate fraction has virtually no sulfur (Table 3-4; Figure 3-6) (Reynolds, 1985). The aromatic fraction has a relatively high concentration of sulfur because low molecular weight thiophenic sulfur compounds act like

Table 3-4. Sulfur concentrations (wt. %) for selected residua separated by the SARA method of fractionation (Reynolds, 1985).

	Fraction			
Feed[a]	Saturate	Aromatic	Resin	Asphaltene[b]
Maya 650°F$^+$ (4.6)	0.04	4.12	5.78	7.80
Arabian Hvy 650°F$^+$ (4.5)	0.01	4.01	6.22	8.23
Beta 650°F$^+$ (4.1)	0.06	3.79	4.60	4.75
Kern River 650°F$^+$ (1.1)	0.01	1.33	1.51	1.51
Hondo 850°F$^+$ (5.9)	0.10	4.37	7.29	7.13
Boscan 1000°F$^+$ (5.6)	0.02	5.41	6.49	7.21

a. (data in parentheses) = sulfur content of residuum, ppm. b. Iso-octane-derived asphaltenes.

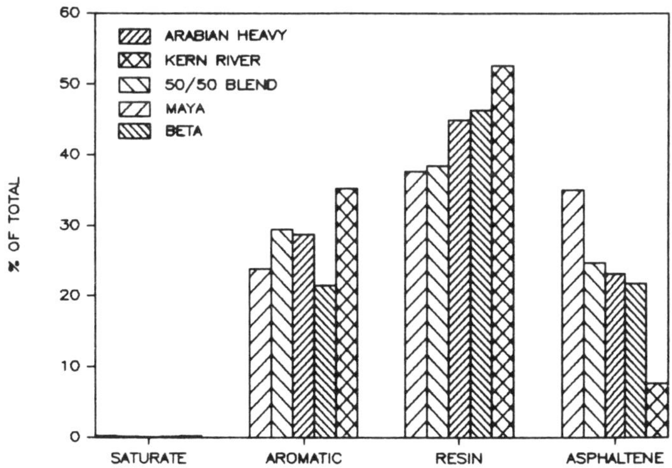

Figure 3-6. Sulfur distribution in the SARA fractions from several; 650°F$^+$ residua (Reynolds, 1985. Copyright Marcel Dekker, Inc. Reproduced with permission).

aromatics. This is a failure of the separation method, because these types of compounds are generally considered part of the polar species in crude oils.

It should also be noted here that the SARA separation suffers from the lack of specific separation of compounds. The fractions are simply operationally defined and do not reflect any specific class or functionality (Chapter 1). This has been a problem in the use of this separation method, especially when *chemical interpretation* of the constituents of these fractions is desired (Speight, 1994).

Even though the asphaltene fractions generally have the highest sulfur content for a specific residuum, the size of the resin fraction often dominates the distribution. Typically, the asphaltenes have the highest concentration of vanadium (Table 3-5) and other metals. Most often, in characterizing metal complexes in crude oils, an asphaltene precipitation is performed that concentrates the vanadium (as well as nickel and iron) (Table 3-5). The nickel and nitrogen distributions are very similar to the vanadium distributions, in general, with the asphaltene fraction having the highest concentrations.

2.4 Heteroatom Distributions in the Acids-Bases-Neutrals Separation

There are many methods other than the SARA separation attempting to simplify, concentrate, isolate, and identify molecular species in crude oils and processed products (Speight, 1991). Some are adaptations of the SARA method

Table 3-5. Vanadium concentrations (ppm) for selected residua separated by the SARA method of fractionation (Reynolds, 1985).

	Fraction			
Feed[a]	Saturate	Aromatic	Resin	Asphaltene[b]
Maya 650°F$^+$ (389)	0	0	242	1602
Arabian Hvy 650°F$^+$ (93)	0	11	81	594
Beta 650°F$^+$ (199)	9	12	147	749
Kern River 650°F$^+$ (48)	0	11	75	361
Hondo 850°F$^+$ (323)	0	2	198	979
Boscan 1000°F$^+$ (1796)	0	0	1595	4410

a. (data in parentheses) = vanadium content of residuum, ppm. b. Iso-octane-derived asphaltenes.

in which further subfractions of SARA fractions are produced (for example, resin-1 and resin-2 are generated by using solvent mixtures to elute resin fractions). Others use completely different approaches such as sequential elution chromatography, which is based on solvent affinity.

One detailed method that avoids using the asphaltene precipitation is the acid-base-neutrals (ABN) separation method (Green et al., 1990, 1994) (Figure 3-7). In this procedure, the crude oil is first separated by molecular distillation into three (or more) fractions, and each fraction is handled differently for further separation. The <345°C (<650°F) fraction is characterized by gas chromatography and by gas chromatography–mass spectrometry methods typically applied to volatile materials. The gas oil fraction (345–540°C; 650–1000°F) is further separated into acidic, basic, and neutral fractions using anion and cation exchange resins. These resins are polystyrene-divinylbenzene type resins with NR_4^+ and sulfonic acid functionalities for the anion and cation resins, respectively. The >540°C (>1000°F) fraction is further separated by a non-aqueous ion exchange procedure using the same resins with different solvent gradients into several ABN fractions. The neutral fractions are also combined for further isolation of sulfides (using a Pd^{II} ligand exchange) and polar neutrals (using silica).

The ABN separation method is extensive and requires more effort than the SARA separation or distillation. However, it produces, in many cases, fractions that are more functionally specific and therefore have more chemical meaning than those produced by the other methods (Table 3-6). One aspect of the separation that is worthy of note is that sulfur compounds occur in every fraction and there is little, if any, partitioning of the sulfur-containing constituents into any one fraction.

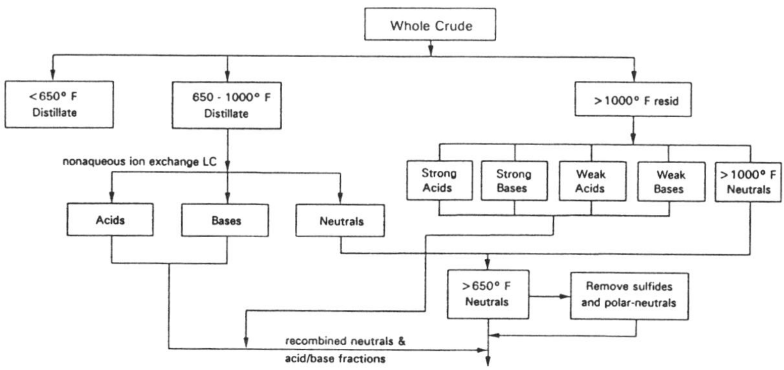

Figure 3-7. The ABN separation method (Green et al., 1994. Copyright American Chemical Society. Reproduced with permission).

Table 3-6. Nitrogen and sulfur concentrations in the ABN fractions of Wilmington 650°F$^+$ residuum (Green et al., 1994).

650-1000°F Distillate

	Distillate	Acids	Bases	Neutrals	Sulfide-Free Neutrals
% of 650°F$^+$	34.3	3.6	1.5	3.2	26.5
S, wt %	1.97	1.23	1.20	2.40	1.40
N, wt %	0.56	2.14	3.90	0.19	0.15

1000°F+ Residuum

	1000°F$^+$	Strong Acids	Weak Acids	Strong Bases	Weak Bases	Neutrals	Sulfide-Free Neutrals
% of 650°F$^+$	65.7	7.4	10.2	7.7	6.2	14.1	21.8
S, wt %	2.66	2.36	2.18	3.05	3.32	2.80	1.58
N, wt %	1.30	1.16	1.60	2.97	2.20	0.58	0.49

It is also important to note that single molecules in many of these fractions contain multiple heteroatoms (and no doubt multiple functionality), making separation into distinct classes difficult. Separation into a fraction containing only sulfur or nitrogen is impossible and, perhaps, meaningless. For the 345-540°C (650-1000°F) fraction, the neutral fraction has the most sulfur-containing constituents, but for the >540°C (>1000°F) residuum, the weak bases fraction exhibits the highest sulfur content. As expected, the majority of the nitrogen-containing components of the feedstock occur in the basic fractions, although nitrogen is found in all of the fractions. Examination of some of these fractions by gas chromatography–mass spectrometry shows phenols, benzothiophenes, indoles, and carbazoles in the basic fractions (Thomson et al., 1994; Green et al., 1994a; Green et al., 1994b).

The appearance of both nitrogen and sulfur in every fraction demonstrates the complexity of the compounds that make up the crude oils. Even the neutral fractions contain some nitrogen. These elements, along with oxygen data, clearly indicate that a large number of multifunctional compounds are found in the gas oil and nondistillable fractions of the feedstock under investigation and, by implication, other crude oils.

Investigation of the concentrations and distributions of vanadium and nitrogen for the Cerro Negro 700°C$^+$ residuum (Table 3-7, Figure 3-8) shows that the vanadium has been divided into porphyrin vanadium (defined as ultraviolet-visible detectable) and nonporphyrin vanadium (total vanadium minus porphyrin

Table 3-7. Distribution of vanadium in the acids-bases-neutral fractions of Cerro Negro 700°C$^+$ residuum (Pearson & Green, 1993).

	Residuum	Strong acids	Weak acids	Strong bases	Weak bases	Neutrals
wt % of total	100	27.9	14.6	10.8	9.5	33.6
Porphyrin V, ppm	97	<22	57	110	220	145
Total V, ppm	1040	1250	1020	1170	602	291
Total Ni, ppm	226	307	273	339	176	78

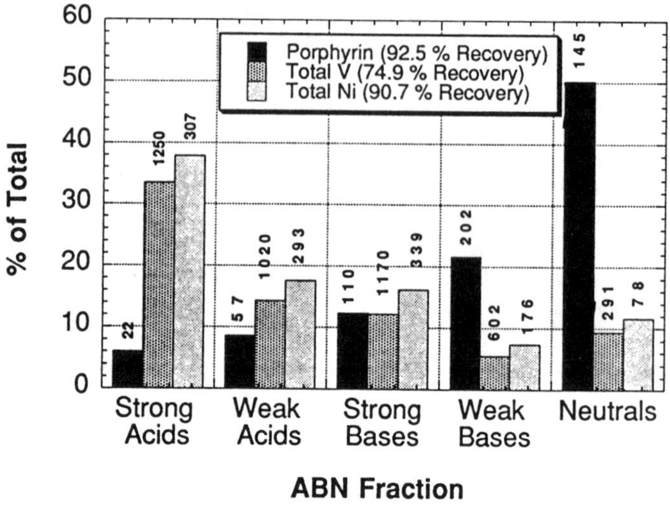

Figure 3-8. Distribution of vanadium, vanadium porphyrins, and nickel in acid-base-neutral fractions of Cerro Negro 700°C^{+} residuum (Pearson & Green, 1993. Copyright American Chemical Society. Reproduced with permission.)

vanadium). For the total vanadium, the strong acids fraction has the highest concentration, while the neutrals has much less. The converse is the case for the porphyrin vanadium.

Porphyrinic vanadium is bound with four nitrogen atoms in a pyrrole environment and is expected to be basic to neutral, as seen in the concentrations. The nonporphyrin vanadium has been shown in other separations to be more polar. The total distribution of nickel is similar to that of vanadium except that the strong base fraction has the highest concentration. For the porphyrinic vanadium, the neutral fraction serves this purpose. The majority of the nickel and vanadium appear in the strong acid fraction, which is inconsistent with the basic nature of porphyrin coordination. This does suggest different types of ligand coordination, as will be discussed in the metals section. However, as indicated by the nitrogen and sulfur (and oxygen) analyses of many types of these fractions, functionality on the perimeter of the porphyrin ring can also change the polarity of the porphyrin greatly (Fish et al., 1984).

2.5 Heteroatom Distribution in Sequential Elution Fractions

Separation of feedstocks by molecular distillation allows deeper distillation cuts than achieved by other distillation methods. However, even with the distillation temperature as high as 705°C (1300°F) atmospheric equivalent boiling point, there remains a residual nondistillable portion of heavy crude oils. This amount can be substantial, 10 to 30% (or more) of heavy crude oils, and the molecular information about this residual material could lead to potentially unlocking secrets of processing. However, the complexity of the residue makes characterization virtually intractable. There have been valiant attempts that have greatly advanced the knowledge of the field. One method developed in an attempt to simplify this material for characterization is the sequential elution fractionation (SEF) method (Altgelt and Boduszynski, 1994).

In this method, the residue is coated onto an inert support composed of a 50/50 mixture of Teflon and glass beads and enough is used to deposit a coating approximately 0.7 mm deep. The material is then packed into a column, and various solvents of increasing polarity are passed over this coated support (*n*-pentane, cyclohexane, toluene, and 4/1 v/v methylene chloride/methanol). This method was designed to separate on the basis of the molecular weight of each compound type, just as distillation does. As a result, the method is considered as an extension of distillation, even though liquid fractionation is employed.

The SEF method shows that the complexity of the crude oil increases in a continuous fashion going from distillate to nondistillable residue, particularly in terms of the number of heteroatoms in a single molecule (Figure 3-4). To some extent, this continuity of structure is in disagreement with the asphaltene concept, which indicates that there is a discrete set of compounds called asphaltenes.

3.0 STRUCTURES OF HETEROATOM FUNCTIONS

The distributions of heteroatoms as a function of boiling point are important from the refining point of view, because of the effects that heteroatoms have on processing. Separation methods alone can be very useful when evaluating a feedstock for processability. Knowledge of the concentrations and structures of these materials in the gas oil and residual fractions is very important for catalyst selection and lifetimes. For example, desulfurization and demetalation of residual materials will probably be done before catalytic cracking. Certain refinery separation methods, such as the solvent deasphalting process, can be evaluated by an asphaltene precipitation, more or less.

From an analytical aspect, the separations above help characterize the type of heteroatomic molecules found in each fraction. For example, from the molecular distillation and the AEBP concept, the heavy fractions are known to contain molecules, on the average, that could contain 10 to 20 sulfur as well as nitrogen atoms. When separation methods are coupled with spectroscopic methods, more information can be elucidated about individual structures and classes of compounds.

The following discussion extracts from the literature some of the proposed and elucidated structures of heteroatomic compounds. Some of these compounds have been detected and unequivocally identified. Some have only been speculated to be the type of compounds that may fit the spectroscopic and analytical evidence. Many represent a class of compounds that includes homologs with various alkyl chain substitutions (homologous series). All the studies done on specific compounds are far too numerous to mention here, and the following discussion is intended to give a flavor of the complexity, variety, and limitations of the types of these compounds in heavy crude oils.

3.1 Structure of Sulfur Compounds

Sulfur concentration is high in many heavy crude oils, residua, and fractions. Examples include 5.3 wt. % sulfur in Boscan crude oil, 5.2 wt. % sulfur in Hondo crude oil, 8.2 wt. % sulfur in the asphaltenes from Arabian heavy crude oil 345°C$^+$ (650°F$^+$) residuum, 7.2 wt. % sulfur in asphaltenes from Boscan crude oil 540°C$^+$ (1000°F$^+$) residuum, 3.3 wt. % sulfur in the weak-base fraction from Cerro Negro 700°C$^+$ residuum, and >7 wt. % sulfur from sequential elution fractions from Boscan crude oil 675°C$^+$ (1250°F$^+$). The major types of sulfur compounds detected have been thiols, sulfides, disulfides, sulfoxides, and thiophenes. Major substitution series have been cyclic, acyclic, alkyl, aryl, arylaryl, and arylalkyl.

Many spectroscopic methods have been used for determining sulfur structures in heavy crude oil fractions. Of particular interest has been the determination of the functional group types in an entire crude or fraction.

Sulfur K-edge x-ray absorption spectroscopy using XANES (x-ray absorption near-edge structure) has been utilized by some researchers to elucidate some very important structural features of the forms of sulfur in heavy crude oils and asphaltenes (George et al., 1989, 1990; Waldo et al., 1991, 1992). One of the essential aspects of the technique is that it can assess the overall concentration of sulfur types, even in nonvolatile materials. The volatility issue is particularly important considering that many detection techniques require some type of volatility for identifying compounds. In addition, many separation techniques combined with spectroscopic examination look at only small portions of the heavy crude oil and often do not reveal an overall picture of the distribution of structural types. The XANES (and EXAFS) technique looks at the entire sample, giving the predominant form of the element.

XANES relies on model compounds, for identification of species in a complex mixture such as heavy crude oil (Figure 3-9). The first coordination sphere around the sulfur atoms is responsible, for the most part, for the particular spectral differences. The third derivative spectra (on the right-hand side) show that two types of compounds can be distinguished, even in a mixture. The distribution of sulfur types (noting that many other model compounds were examined and tested) determined from this study was (% sulfide, % thiophenic): asphaltene 1: 0, 100; asphaltene 2: 43, 37; asphaltene 3: 50, 50; Rasa coal: 30, 70. Results from another XANES study that examined different samples (Table 3-8) indicated the presence of sulfoxide functions as well as sulfide and thiophenic functions. In these samples, the majority of the sulfur is in the thiophenic form. The Uinta basin sample (Table 3-8) is a shale and shows that the sulfoxide form is greatly enhanced by weathering, consistent with the oxidation of sulfur. Note also that the asphaltene separation does little to partition the sulfur types, once again substantiating that crude oil is a continuum.

Techniques employing volatility, such as gas chromatography–mass spectrometry (GC-MS) and liquid chromatography–mass spectrometry (LC-MS), have been used to identify a huge number of sulfur-containing compounds in various fossil fuels (Orr and Sinninghe Damste, 1990) (Figures 3-10 and 3-11) and provide an indication of the variety of these types of compounds in fossil fuels: thiols, disulfides, thiaalkanes and thiacycloalkanes, thiophenes, and condensed thiophenes. A variety of biomarker-type materials have also been detected and isolated, and in some cases spectroscopic evidence suggests a variety of structures (Figure 3-12).

3.2 Structure of Nitrogen Compounds

Like sulfur, nitrogen is found in many heavy crude oils, residua, and fractions. The concentration tends to be a great deal lower in many cases than the corresponding sulfur concentration, although the crude oils from China (Table 3-1) are examples in which this is not true. Examples include Boscan crude oil: 0.64 wt. %; Hondo crude oil: 0.62 wt. %; asphaltenes from Arabian heavy crude oil 345°C$^+$ (650°F$^+$) residuum: 0.15 wt. %; asphaltenes from Boscan crude oil 540°C$^+$ (1000°F$^+$) residuum: 0.65 wt. %; the strong base fraction from Cerro Negro crude oil 700°C$^+$ residuum: 2.97 wt. %.

The different types of nitrogen detected are basic nitrogen compounds (strong bases): pyridines, amines, alkylarylamines, and *N*-alkylindoles; neutral

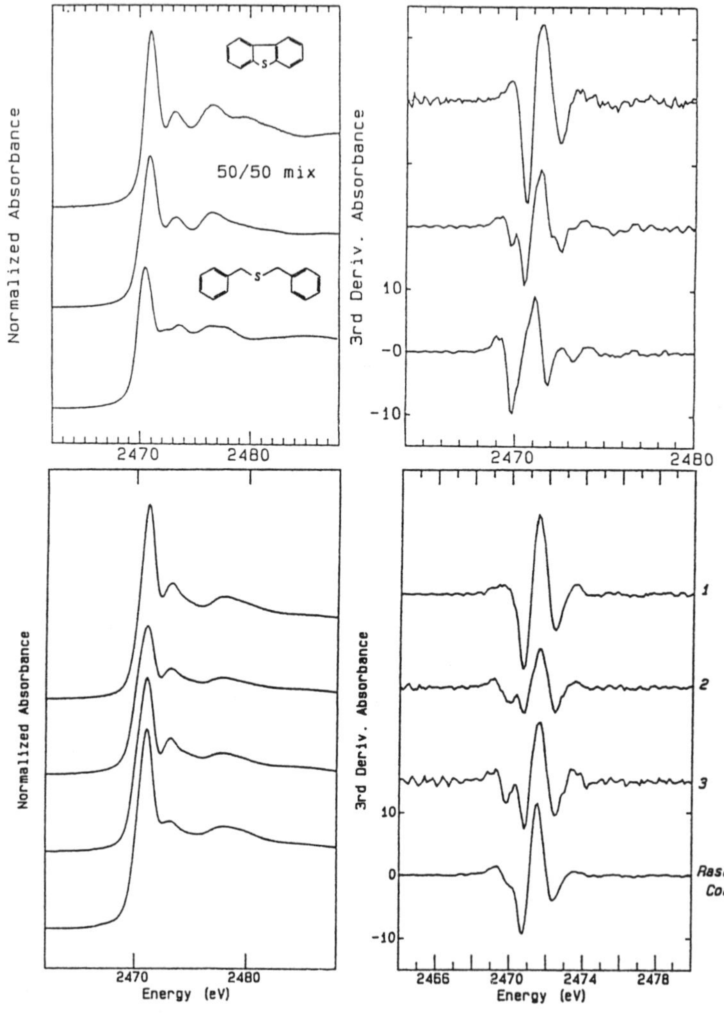

Figure 3-9. Sulfur K-edge x-ray absorption spectrum using x-ray absorption near edge structures (XANES) of (top) sulfur model compounds and (bottom) three asphaltenes and Raza coal (George et al., 1990. Copyright American Chemical Society. Reproduced with permission).

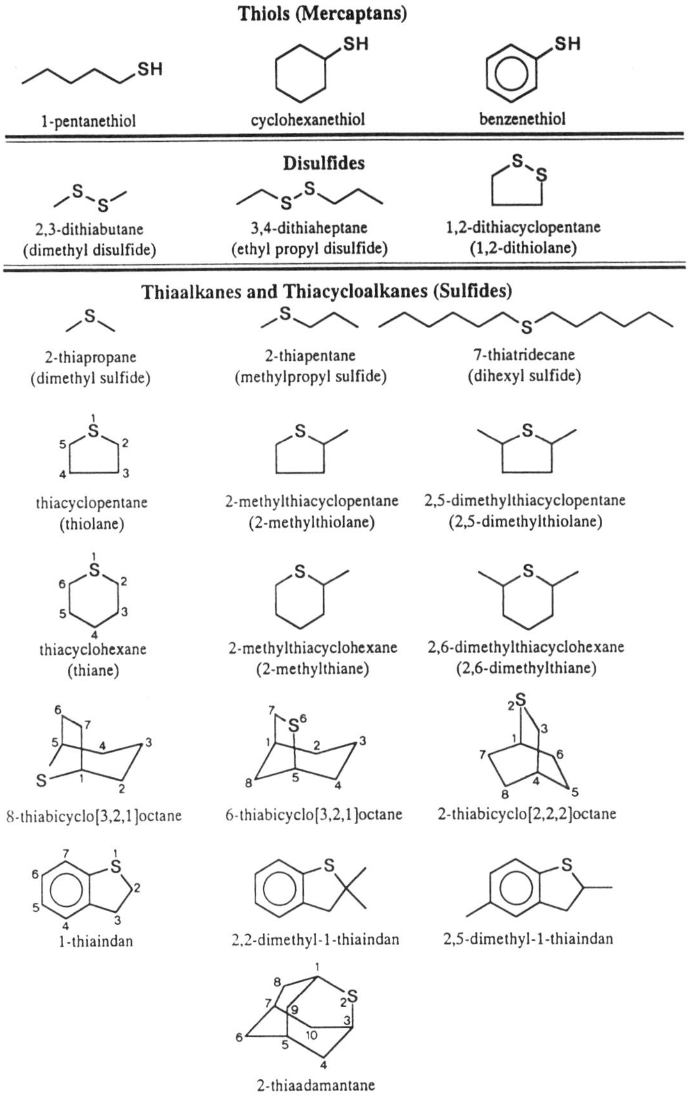

Figure 3-10. Sulfur compounds found in fossil fuels (Orr & Sinninghe Damste, 1990. Copyright American Chemical Society. Reproduced with permission).

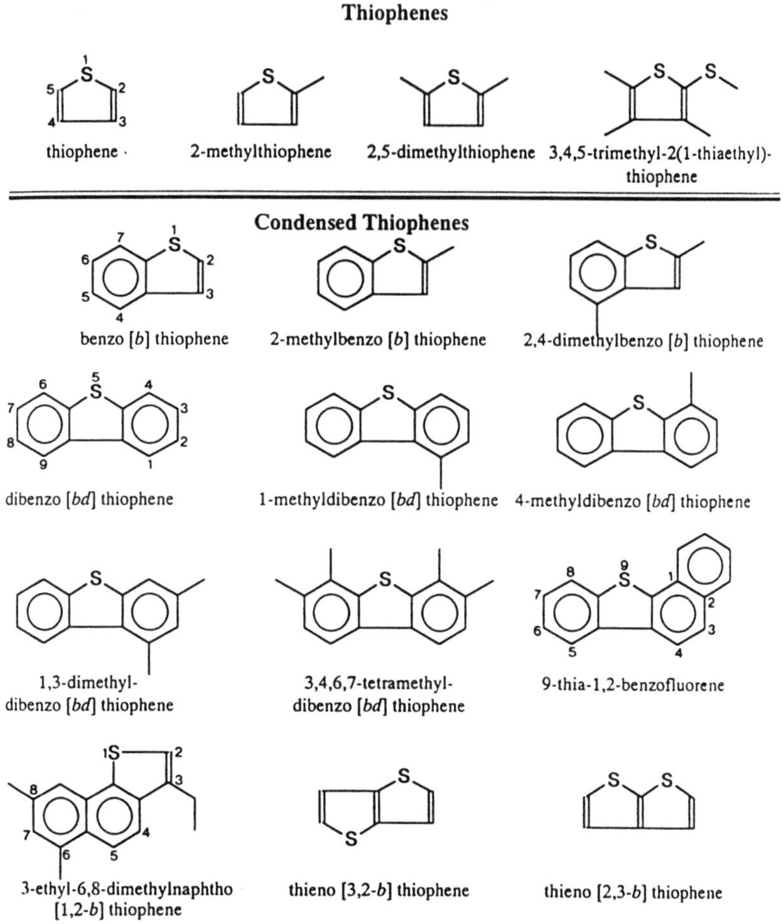

Figure 3-11. Thiophene-type compounds found in fossil fuels (Orr & Sinninghe Damste, 1990. Copyright American Chemical Society. Reproduced with permission).

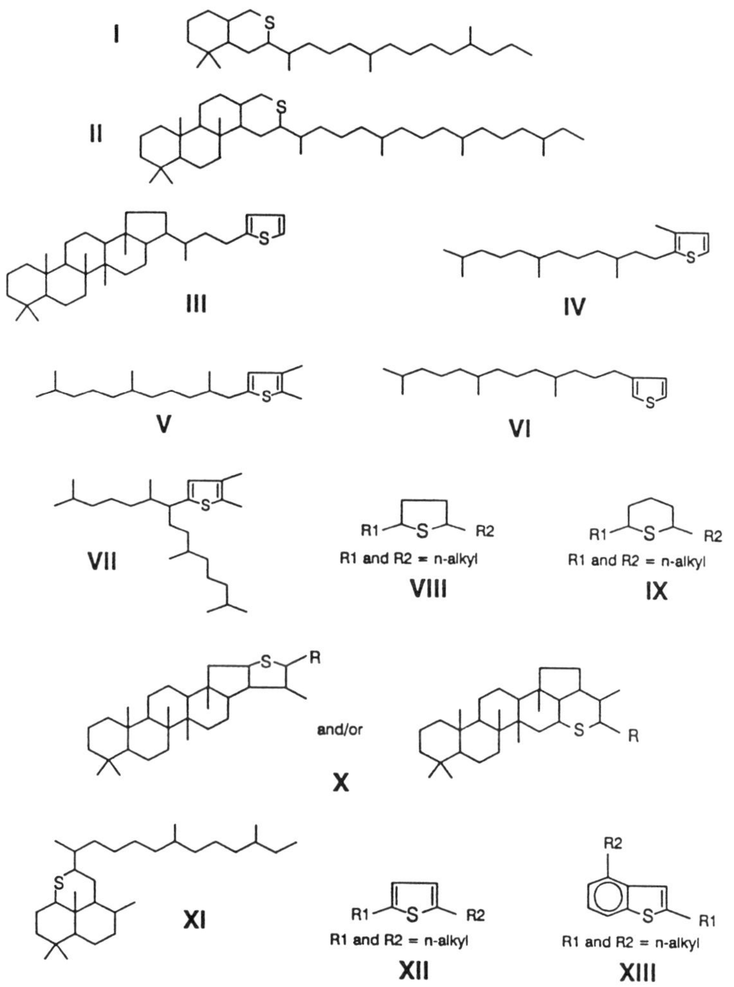

Figure 3-12. Sulfur-containing biomarkers found in fossil fuels (Orr & Sinninghe Damste, 1990. Copyright American Chemical Society. Reproduced with permission).

Table 3-8. XANES analysis of selected samples (Waldo et al., 1991).

Sample	Total S wt %	% Sulfide	% Thiophene	% Sulfoxide
Monterey (CA)	8.00	38	60	2
Maltenes	8.79	39	60	1
Asphaltenes	9.48	31	67	2
Monterey	4.65	33	57	10
Monterey	3.70	31	63	6
Monterey	2.22	40	56	4
Kimmeridge (North Sea)	0.54	9	82	5
Rozel Point (Utah)	13.47	38	59	3
Maltenes	13.31	41	58	1
Asphaltenes	14.50	34	63	2
Uinta Basin (Utah)	0.21	23	51	26
Air for 1 year	0.21	8	55	37
Uinta Basin (Utah)	0.65	30	65	5
Shengli (China)	0.98	22	73	5

nitrogen types (weak bases): alkyl-indoles and acridines; and acidic nitrogen: indoles, carbazoles, porphyrins, and amides.

The distribution of nitrogen types in various acid and base fractions from several distillate fractions from heavy crude oils (345–540°C, 650–1000°F, and 540–675°C, 1000–1250°F) has been studied (McKay et al., 1975, 1976) by infrared spectroscopy and titration (Table 3-9). Note that the pyridine derivatives tend to be in the highest concentration for a specific distillate fraction, followed by diaza compounds. Occasionally, pyrroles and amides are found in high concentrations.

The acid-base-neutral fractions containing high nitrogen concentrations have been examined in much more detail by field ionization mass spectrometry (Boduszynski, 1988) (Figure 3-13), which indicates that multiring aromatic nitrogen heterocyclic compounds are present in the basic nitrogen fraction (basic alumina back-flushed with methylene chloride/methanol) as well as substituted quinolines and multiring pyrrolic compounds in the pyrrolic fraction (acidic alumina eluted with methylene chloride). Clearly, the nitrogen compounds stand out as homologous series. However, the intensity at all masses (Figure 3-13)

Table 3-9. Distribution (wt. % distillate) of nitrogen compounds (determined by infrared spectrometry) in acid and base fractions of 370 to 535°C and 535 to 675°C atmospheric equivalent boiling range fractions of selected residua (McKay et al., 1975; table from Altgelt and Boduszynski, 1994).

Crude oil boiling range	Pyridine benzologs B	Diaza compounds B	Pyrrole derivatives B	Pyrrole derivatives A	Amides B	Amides A
Wilmington						
370–535°C	3.5	1.16	0.5	0.9	1.7	0.12
535–675°C	7.2	2.9	0.5	–	–	0.12
Gach Saran						
370–535°C	1.2	0.54	0.04	0.53	0.27	0.34
535–675°C	3.9	2.6	0.17	–	2.0	–
Recluse						
370–535°C	0.5	0.37	0.02	1.0	0.20	0.05
535–675°C	1.9	0.86	0.03	1.7	0.53	0.24
Swan Hill						
370–535°C	0.7	0.57	0.07	0.8	0.81	0.29
535–675°C	2.2	1.31	0.18	–	0.81	–

indicates that even after separation the fractions are incredibly complex and that many other species can be present, including ones containing sulfur and oxygen.

XANES has also been used to determine nitrogen structures (Mitra-Kirtley, Mullins, Branthaver, et al., 1993, Mitra-Kirtley, Mullins, van Elp, et al., 1993) in several *n*-heptane asphaltenes, kerogen (mineral free by HCl and HF deashing), and corresponding extracted bitumen (benzene/methanol extraction) (Table 3-10). The kerogens were from the Green River formation and New Albany formation. A very old kerogen, demineralized and stored in air for many years, also was examined. As in the XANES study of sulfur, the technique relies on model compounds and measures overall predominant forms. The pyridone and aromatic amine model compounds were not used in the asphaltene study.

Pyrrole derivatives are found in the highest concentration in the asphaltenes, kerogen, and bitumen. Pyridine is the functionality found in the second highest concentration in most cases. Pyridone functionality is observed in

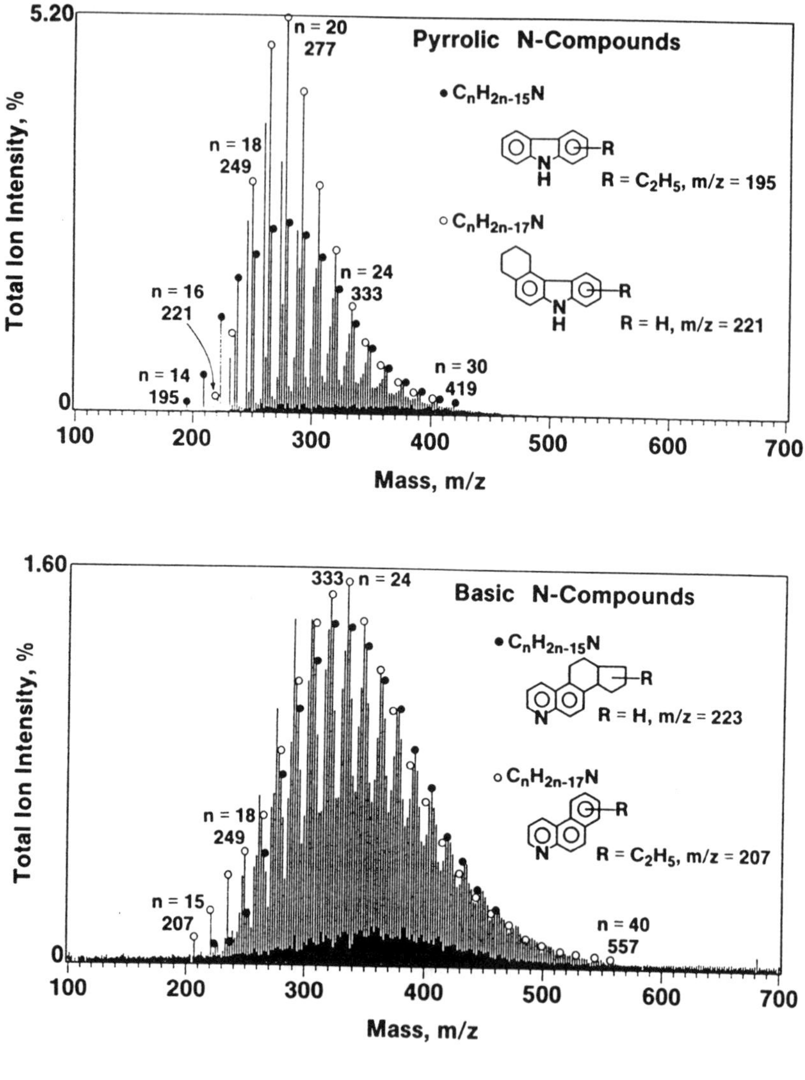

Figure 3-13. Distribution of nitrogen compounds in Kern River vacuum gas oil (Boduszynski, 1988).

Table 3-10. Distribution of nitrogen structural types by XANES in asphaltenes, kerogens, and bitumens (Mitra-Kirtley, Mullins, Branthaver et al., 1993; Mitra-Kirtley, Mullins, van Elp et al., 1993). (Pyridone and aromatic amine model compounds were not used in the asphaltene study.)

Sample N	% pyridine	% pyridone	% pyrrole	% aromatic amines	% saturated amines	wt %
Canadian Asphaltenes	26	--	74	--	0	0.91
Kuwait 3 Asphaltenes	22	--	77	--	2	0.95
Kuwait 2 Asphaltenes	22	--	79	--	0	0.87
California Asphaltenes	30	--	65	--	2	2.32
France Asphaltenes	22	--	76	--	3	0.54
Kuwait 1 Asphaltenes	2	--	87	--	2	0.98
Green River Kerogen 1	20	23	42	15	0	1.09
Green River Bitumen 1	27	11	49	11	2	0.83
Green River Kerogen 2	34	11	42	2	11	1.03
Green River Bitumen 2	20	19	49	12	0	1.47
Weathered Kerogen	24	20	45	10	0	1.81
New Albany Bitumen	14	11	67	8	0	2.60

all kerogen and bitumen at reasonably high concentrations as well as aromatic amine derivatives. There is little evidence for saturated amines.

Not as many studies to determine specific structures of nitrogen compounds are found in the literature, although some selected structures (Figures 3-14 and 3-15) have either been detected or postulated and represent many of the types of nitrogen compounds (McKay et al., 1976; Sturm et al., 1987; Thomson et al., 1994).

Z-Series[a] Designation	Homologous Series Compound Type Name	Typical Compound[b] Structures
-15N	Alkylcarbazoles	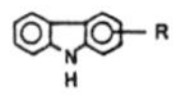
-17N	Alkylnaphthenocarbazoles and/or alkylcyclohexylcarbazoles	
-19N	Alkyldinaphthenocarbazoles	
-21N	Alkylbenzocarbazoles	
-23N	Alkylphenylcarbazoles and/or alkylnaphthenobenzocarbazoles	
-25N	Alkyldinaphthenobenzocarbazoles	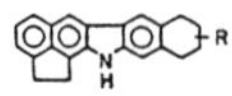
-27N	Alkyldibenzocarbazoles	

Figure 3-14. Nitrogen compounds in crude oils and processed products (Sturm et al., 1987. Copyright American Chemical Society. Reproduced with permission).

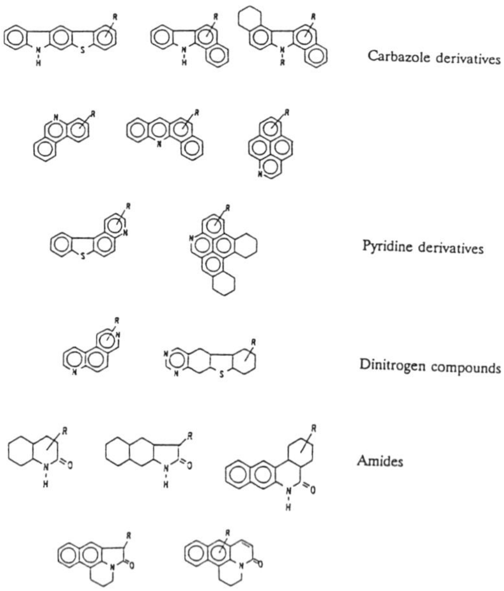

Figure 3-15. Nitrogen compounds in crude oils and processed products (McKay et al., 1976. Copyright American Chemical Society. Reproduced with permission).

3.3 Structure of Metal Compounds

Relative to sulfur and nitrogen, vanadium and nickel are found in much lower concentrations in heavy crude oils, residua, and fractions. That does not reduce their impact in processing, however. In the previous sections, it was mentioned that the studies on distillation show that the metals are concentrated in the nondistillable portion. Not until the distillation temperatures are above 540°C (1000°F) atmospheric equivalent boiling point do the metals begin to volatilize. In most separation studies, the metals are also closely aligned to polar fractions. Also, asphaltenes tend to have higher concentrations than maltenes.

Examples include Boscan crude oil: 920 ppm vanadium, 90 ppm nickel; Maya crude oil: 242 ppm vanadium, 39 ppm nickel; asphaltenes from Maya 345°C (650°F$^+$): 1602 ppm vanadium, 286 ppm nickel; asphaltenes from Boscan 540°C (1000°F$^+$): 4410 ppm vanadium, 375 ppm nickel; the strong acid fraction from Cerro Negro crude oil 410°C (770°F$^+$): 120 ppm vanadium, 307 ppm nickel; a sequential elution fraction from Boscan crude oil 705°C (1300°F$^+$) residuum: 4400 ppm vanadium, 400 ppm nickel.

The metals are found bound in two types, metallo-petroporphyrins and metallo-nonporphyrins. The former have the distinctive ultraviolet-visible Soret band due to the highly conjugated porphyrin ring system, and the latter is comprised of whatever metal concentrations are not accountable by the ultraviolet-visible Soret band.

An enormous amount of analytical work has been performed on the metallo-petroporphyrins since their discovery in fossil fuels in 1935 (Treibs, 1935). Because of the distinctive spectroscopic properties of the porphyrins, many studies have isolated specific structural types. Hundreds of structures of different porphyrins have been determined by mass spectrometry for a variety of fossil fuel materials (Figure 3-16) (Quirke, 1987). The porphyrin ring is essentially the same in all these cases and the molecular differences are reflected in the substitution patterns produced by the attendant side positions. These substitutions include alkyl, aryl, and cyclic and associated homologous series.

Porphyrins do not account for all the vanadium compounds in the crude oil or fraction. Division of the metals content based on the Soret band of the porphyrins can be made (Table 3-11) (Quann et al., 1988). Clearly, not all the vanadium can be accounted for as porphyrin, even in separated fractions.

Other data derived from size exclusion chromatography profiles using vanadium specific detection by inductively coupled plasma atomic emission spectroscopy (V-SEC-ICP-AES) of Morichal heavy crude oil from Venezuela before and after porphyrin extraction (methanol extraction of cellulose coated with residuum) (Reynolds and Biggs, 1988) emphasize the controversy (Figure 3-17).

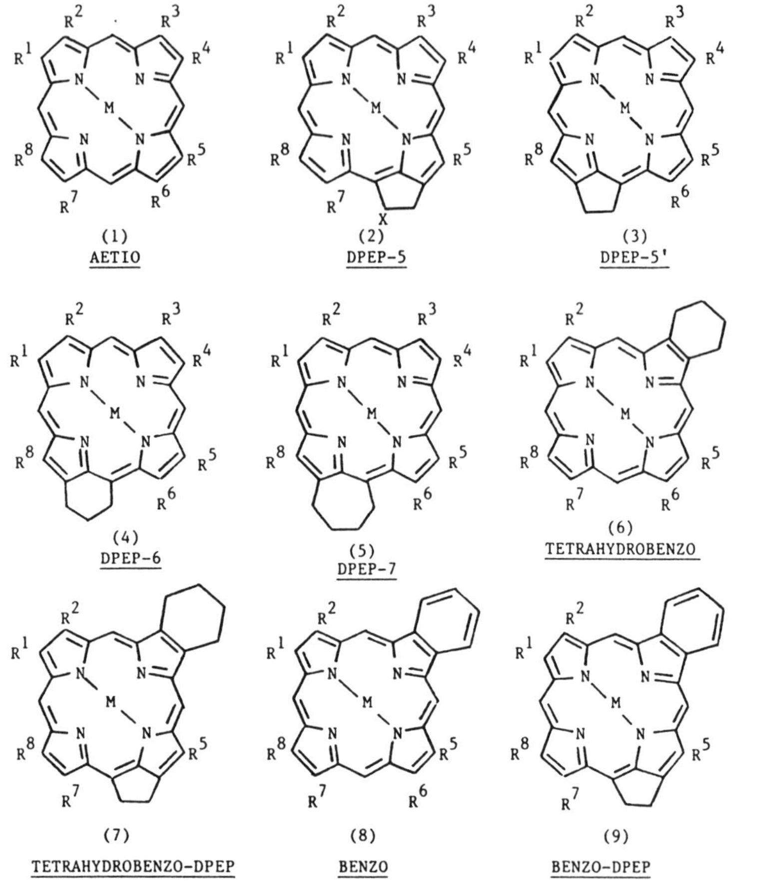

1(a) $R^1,R^3,R^5,R^8 = CH_3$; $R^2,R^4,R^6,R^7 = C_2H_5$

1(b) $R^1,R^3,R^5,R^8 = CH_3$; $R^2,R^4,R^7 = C_2H_5$; $R^6 = H$

2(a) $R^1,R^3,R^5,R^8 = CH_3$; $R^2,R^4,R^7 = C_2H_5$; $X = OH$

2(b) $R^1,R^3,R^5,R^8 = CH_3$; $R^2,R^4,R^7 = C_2H_5$; $X = H$

2(c) $R^1,R^3,R^5,R^8 = CH_3$; $R^2,R^4 = C_2H_5$; $R^7 = CH_2CH_2CO_2H$; $X = H$

Figure 3-16. Structure of metallo-porphyrins found in fossil fuels (Quirke, 1988. Copyright American Chemical Society. Reproduced with permission).

Table 3-11. Concentrations (ppm) of vanadium as prophyrins and nonporphyrin fractions for four crude oils (Quann et al., 1988).

	Fraction	Boscan	La Luna	Baxterville	Belridge
Present as	Crude Oil	350	56	46.7	17
porphyrins	Asphaltenes	868	200	230	180
	Resins	622	460	73	31
	Oil	21.6	9.05	0.56	2.7
Present as	Crude oil	784	152	~0	15
nonporphyrins	Asphaltenes	3612	2740	~0	95
	Resins	439	370	~0	58.7
	Oil	9.8	3.3	~0	~0

The residuum exhibits vanadium in a wide size range (size decreases with elution time). Extraction, which accounts for 96% of the porphyrins detected in the ultraviolet-visible, shows a narrow profile in the small molecular size range. The residual material, which contains the bulk of the vanadium, does not have an ultraviolet-visible Soret band and spans the entire size range of the crude oil.

The different types of metals (Figure 3-17) led to some controversy because only metallo-petroporphyrins have been unequivocally identified. This had led to many studies over the years which indicated that the metallo-nonporphyrin compounds are vanadium bound with four nitrogen atoms in the first coordination sphere as porphyrins or with other heteroatoms (combinations of nitrogen, sulfur, and oxygen) not in the porphyrin environment.

EXAFS studies (Goulon et al., 1984) exemplify the position that all the vanadium complexes in crude oils are porphyrin-like (Figure 3-18); clearly, the spectra overlap to some extent, although deviations are evident around an R/Å of 3. The interpretation of these results is that the major and probably only binding around the vanadium is due to the porphyrin four nitrogen atom system. Reduced porphyrins such as chlorins and corrins could be possible. However, these are still basically the porphyrin coordination sphere. These typical EXAFS results led many researchers in the field to believe that all the vanadium is bound as porphyrin.

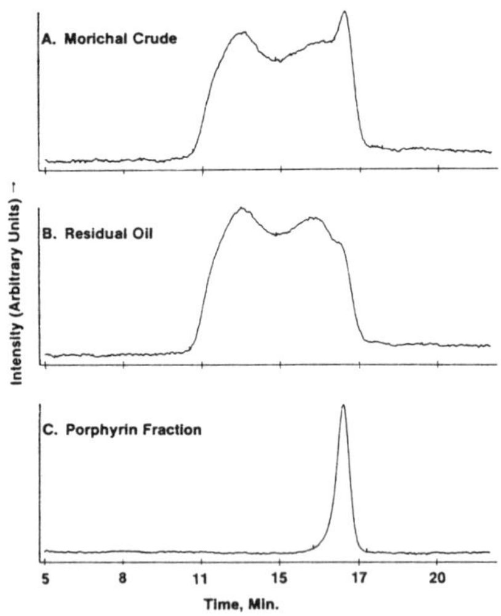

Figure 3-17. Vanadium-SEC-ICP-AES profiles for Morichal crude oil, isolated porphyrin fraction, and residual material from the fraction (Reynolds and Biggs, 1988. Copyright American Chemical Society. Reproduced with permission).

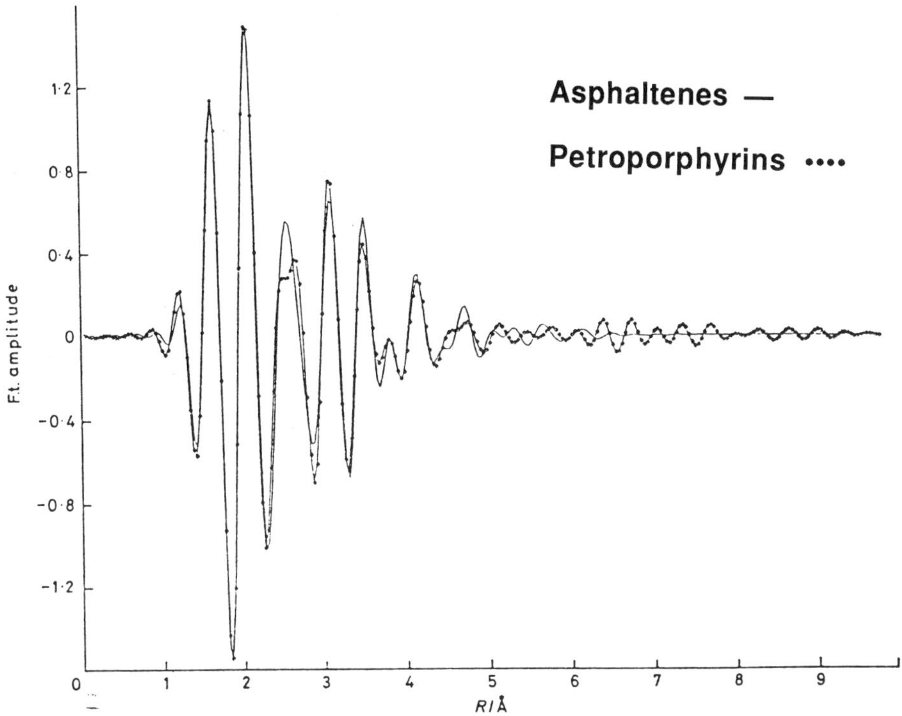

Figure 3-18. EXAFS of Boscan asphaltenes and isolated petroporphyrins (Goulon et al., 1984. Copyright The Chemical Society of London. Reproduced with permission).

Other evidence indicates that the metallo-nonporphyrins are not porphyrin bound (Reynolds et al., 1987; Sundaraman et al., 1988). A processing study (Figure 3-19) (Pearson and Green, 1993) coupled with extensive acid-base-neutral separation provides evidence that the nonporphyrins cannot easily be discounted as simply porphyrins that lack the ultraviolet-visible spectral properties for some reason. Cerro Negro 700°C^{+} residuum was hydrotreated at various severities. The severity was varied primarily by changing temperature, although pressure and residence time were also changed.

The feed has approximately 10% of the vanadium bound as petroporphyrin and 90% as metallo-nonporphyrin. At the lowest severity, a small amount of the porphyrins is removed, while the total metals remain the same. Various researchers argue that the metallo-nonporphyrins are really porphyrins but the ultraviolet-visible spectrum is altered by agglomeration and/or burial in a tertiary structure, reducing the intensity of the Soret band. Detailed investigation of the hydrotreated products has shown that hydrotreating reduces this agglomeration or tertiary structure (Reynolds and Biggs, 1988), so mild processing should increase the amount of porphyrins. However, the behavior of the product at 300°C and 230 psig for 60 minutes is opposite to this: the amount of porphyrins decreases and the total metals remain the same, although there is a shift to the same molecular size as the porphyrins. This is even more dramatically shown in the 350°C and 2700 psig for 60 minutes product. At the higher severities, demetalation becomes facile, and both porphyrins and nonporphyrins are removed.

Hydrotreating of specific acid-base-neutral fractions that were separated from various residua also provided valuable information (Table 3-12). There was no evidence that any appreciable amount of porphyrins was produced. The total demetalation was very low, particularly for the Laganitas and Merey Compo residua.

Many structures have been proposed for the metallo-nonporphyrins (Figure 3-20) (Yen, 1975). Summed up, these studies show the following. The ultraviolet-visible spectral quantitation of many oils accounts for only up to 50% of the vanadium is bound as porphyrin. Chromatographic studies have produced

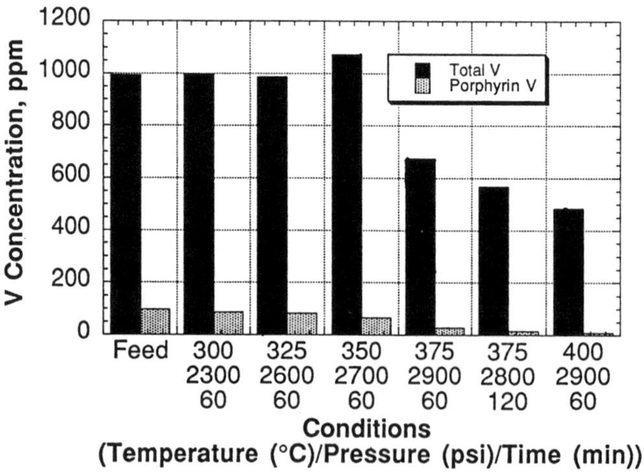

Figure 3-19. Hydrotreatment of Cerro Negro 700°C$^+$ residuum at various severities (data from Pearson and Green, 1993. Copyright American Chemical Society. Reproduced with permission).

Table 3-12. Hydrotreatment of various acids-base-neutral fractions at 350°C, 1200 psi (Pearson and Green, 1993).

Sample	V (ppm)	Ni (ppm)	Porphyrin V (ppm)
CN 700°C$^+$ strong base feed	1240	339	<9
CN 700°C$^+$ strong base product	1100	278	10
CN 700°C$^+$ strong acid feed	1260	286	<9
CN 700°C$^+$ strong acid product	920	225	<10
Lag 510°C$^+$ stong acid feed	1500	149	<10
Lag 510°C$^+$ stong acid product	1525	189	<8
Merey 510°C$^+$ stong acid feed	1120	276	<12
Merey 510°C$^+$ stong acid product	1115	234	17
Wlmgtn 535°C$^+$ stong acid feed	157	209	<9
Wlmgtn 535°C$^+$ stong acid product	124	205	18
Maya 500°C$^+$ stong acid feed	1100	247	<9
Maya 500°C$^+$ stong acid product	906	197	20

Lag = Laganitas, Merey = Merey Compo, Wlmgtn = Wilmington.

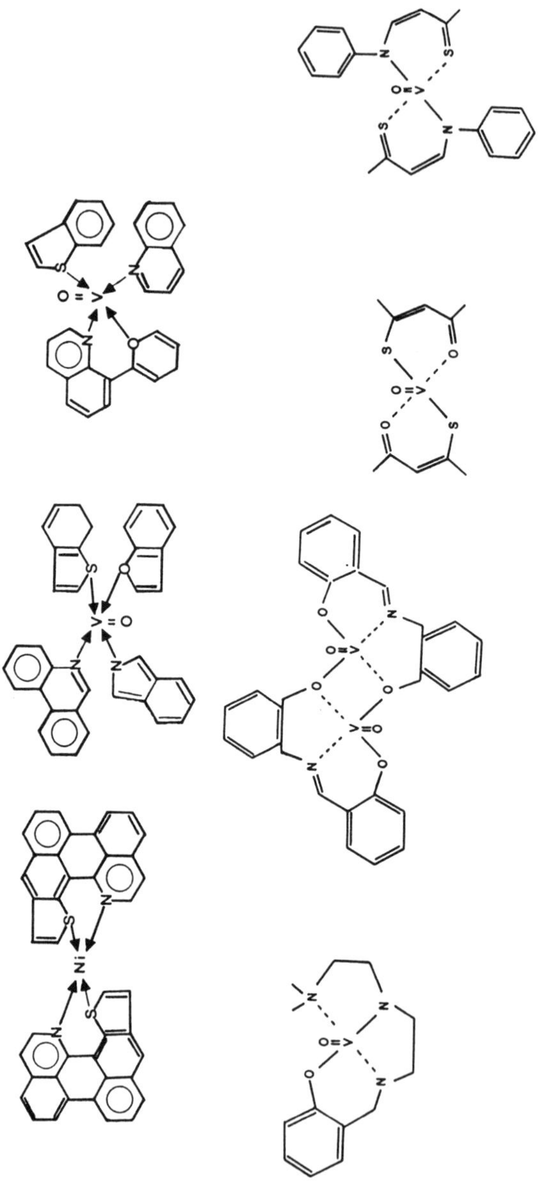

Figure 3-20. Models for metallo-nonporphyrin coordination (Yen, 1975. Copyright Ann Arbor Scientific. Reproduced with permission.).

Table 3-13. Average number of heteroatoms per molecule in sequential elution fractions (Altgelt & Boduszynski, 1994).

Fraction	Mid-AEBP (°F)	Cum. wt% from AR	Mol. wt.[a]	"Z"[b]	Average number of atoms/molecule								
					C_t	C_{ar}	H	S	N	O	V	Ni	Fe
Kern River													
SEF-1	1365	90.3	1110	−44	79	29	114	0.462	1.112	0.737	0.001	0.002	0.001
SEF-2	1945	96.0	2520	−141	179	80	218	1.141	3.957	1.888	0.009	0.019	0.013
SEF-3	2250	99.1	3620	−216	225	138	294	1.617	6.204	2.986	0.019	0.025	0.027
SEF-4	—	99.6	—	—	—	—	—	—	—	—	—	—	—
California Offshore													
SEF-1	1660	78.4	1910	−67	131	35	195	3.49	1.26	0.74	0.006	0.002	0.001
SEF-2	2305	87.9	4150	−203	282	110	362	8.25	5.18	2.18	0.076	0.017	0.012
SEF-3	2670	98.6	6070	−338	413	181	488	12.7	8.88	4.55	0.168	0.038	0.033
SEF-4	—	99.7	—	—	—	—	—	—	—	—	—	—	—
Arabian Heavy													
SEF-1	1565	87.7	1615	−60	112	36	163.7	2.745	0.461	—	0.002	0.0004	—
SEF-2	2310	95.9	3870	−224	261	134	298.7	9.155	2.460	—	0.034	0.009	—
SEF-3	2570	99.0	5060	−313	346	192	378.3	12.723	3.617	—	0.051	0.016	—
SEF-4	—	99.5	—	—	—	—	—	—	—	—	—	—	—
Maya													
SEF-1	1530	78.9	1520	−56	105	35	153	2.35	0.49	—	0.003	0.0008	—
SEF-2	2070	85.5	2970	−158	199	96	240	6.28	2.25	—	0.046	0.009	—
SEF-3	2580	98.5	5170	−315	351	192	387	11.62	4.98	—	0.188	0.029	—
SEF-4	—	99.8	—	—	—	—	—	—	—	—	—	—	—
Boscan													
SEF-1	1510	74.2	1420	−57	96	30	135.9	2.718	0.639	—	0.019	0.002	—
SEF-2	2160	87.3	3460	−175	231	103	287.0	7.590	3.806	—	0.223	0.018	—
SEF-3	2590	98.0	5510	−307	370	192	432.0	12.230	7.087	—	0.479	0.037	—
SEF-4	—	100.0	—	—	—	—	—	—	—	—	—	—	—

[a]Number-average molecular weight determined by VPO in pyridine. [b]"Z" in the general formula $C_nH_{2n+Z}X$

many isolated porphyrin fractions but also fractions in which the vanadium does not have porphyrin characteristics. Processing studies show metal-containing products that do not have petroporphyrin characteristics as precursors. No metallo-nonporphyrin complexes have been isolated and unequivocally identified. EXAFS and XANES studies show that the dominant form of vanadium has four nitrogen atoms in the first coordination sphere, with a porphyrin-like secondary shell.

4.0 CONCLUDING REMARKS

Following on from these studies, it is possible to make deductions in terms of the average number of heteroatoms per molecule in sequential elution fractions of several heavy crude oils (Table 3-13). Based on the atmospheric equivalent boiling point concept, these fractions are effectively the highest boiling range fractions ever separated and analyzed. The average molecule of a sequential elution fraction can have well over 20 heteroatoms in it. Chances are, in a specific molecule, that the heteroatoms are arrange in different functionalities, making an incredibly complex molecule. Considering how many different combinations are possible, the chances of determining every structure in the mixture are very low. Because of this seemingly insurmountable task, it may be better to spend the investment in finding ways of utilizing this material economically, rather than exactly what the material is.

5.0 REFERENCES

Ali, M. F., Hasan, M., Bukhari, A., and Saleem, M. 1985. Oil & Gas J. August 12: 71.

Altgelt, K. H., and Boduszynski, M. M. 1992. Energy and Fuels. 6: 68.

Altgelt, K. H., and Boduszynski, M. M. 1994. Compositional Analysis of Heavy Petroleum Fractions. Marcel Dekker Inc., New York.

Boduszynski, M. M. 1987. Energy and Fuels. 1: 2.

Boduszynski, M. M. 1988. Energy and Fuels. 2: 597.

Boduszynski, M. M., and Altgelt, K. H. 1992. Energy and Fuels. 6:72.

Chrones, J., and Germain, R. R. 1989. Fuel Sci. Technol. Int. 7: 783.

Dolbear, G., Tang, A., and Moorehead, E. L. 1987. Symposium Series. No. 344. Amercian Chemical Society, Washington, DC. p. 221.

Fish, R. H., Komlenic, J. J., and Wines, B. K. 1984. Anal. Chem. 56: 2542.

George, G. N., and Gorbaty, M. L. 1989. J. Am. Chem. Soc. 111: 3182.

George, G.N., Gorbaty, M.L., and Kelemen, S.R. 1990. Symposium Series No. 429. Amercian Chemical Society, Washington, DC. p. 220.
Goulon, J., Retournard, A., Friant, P., Goulon-Ginet, C., Berthe, C., Muller, J-F., Poncet, J-L., Guilard, R., Escalier, J-C., and Neff, B. 1984. J. Chem. Soc. Dalton Trans. 1095.
Green, J. A., Green, J. B., Grigsby, R. D., Pearson, C. D., Reynolds, J. W., Shay, J. Y., Strurm, G. P. Jr., Thomson, J. S., Vogh, J. W., Vrana, R. P., Yu, S. K-T., Diehl, B. H., Grizzle, P. L., Hirsch, D. E., Hornung, K. W., Tang, S-Y., Carbognani, L., Hazos, M., and Sanchez, V. 1990. NIPER Report No. 452. Volumes 1 and 2. Bartlesville, Oklahoma.
Green, J. B., Yu, S. K-T., and Varna, R. P. 1994a. J. High Resolution Chromatogr. 17: 427.
Green, J. B., Yu, S. K-T., and Varna, R. P. 1994b. J. High Resolution Chromatogr. 17: 439.
Green, J. B., Zagulka, E. J., Reynolds, J. W., Wandke, H. H., Young, L. L., and Chew, H. 1994. Energy and Fuels. 8: 856.
Liang, W., Que, G., and Chen, Y. 1991. Proceedings. 1990. Eastern Oil Shale Symposium. p. 24 and p. 90.
McKay, J. F., Cogswell, T. E., Weber, J. H., and Latham, D. R. 1975. Fuel. 54: 50.
McKay, J. F., Weber, J. H., and Latham, D. R. 1976. Anal. Chem. 48: 891.
Mitra-Kirtley, S., Mullins, O. C., Branthaver, J. F., and Cramer, S. P. 1993. Energy and Fuels. 7: 1128.
Mitra-Kirtley, S., Mullins, O. C., van Elp, J., George, S. J., Chen, J., and Cramer, S. P. 1993. J. Am. Chem. Soc. 115: 252.
Orr, W. L., and Sinninghe Damste, J. S. 1990. Symposium Series No. 429. Amercian Chemical Society, Washington, DC. p. 2.
Pearson, C. D., and Green, J. B. 1993. Energy and Fuels. 7: 338.
Quann, R. J., Ware, R. A., Hung, C.-W., and Wei, J. 1988. Adv. Chem. Eng. 14: 95.
Quirke, J. M. E. 1987. Symposium Series No. 344. American Chemical Society, Washington, DC. p. 308.
Reynolds, J. G. 1985. Liquid Fuels Technol. 3: 73.
Reynolds, J. G. 1996. Unpublished data.
Reynolds, J. G., and Biggs, W. R. 1986. Fuel Sci. Technol. Int. 4: 749.
Reynolds, J. G., and Biggs, W. R. 1988. Accounts Chem. Res. 21: 319.
Reynolds, J. G., Biggs, W. R., and Bezman, S. A. 1987. Symposium Series No. 344. American Chemical Society, Washington, DC. p. 205.
Speight, J. G. 1991. The Chemistry and Technology of Petroleum. 2nd Edition. Marcel Dekker Inc., New York.

Speight, J. G. 1994. In Asphaltenes and Asphalts, 1. T.F. Yen and G.V. Chilingarian (Editors). Elsevier, Amsterdam, The Netherlands. Chapter 2.

Speight, J. G., Long, R. B., and Trowbridge, T. D. 1984. Fuel., 63: 616.

Sturm, G. P. Jr., Green, J. B., Tang, S. Y., Reynolds, J. W., and Yu, S. K-T. 1987. Preprints Div. Petrol. Chem. Am. Chem. Soc. 32(2): 369.

Sundararaman, P., Biggs, W. R., Reynolds, J. G., and Fetzer, J. C. 1988. Geochim. Cosmochim. Acta. 52: 2337.

Thomson, J. S., Green, J. B., McWilliams, T. B., and Yu, S. K-T. 1994. High Resolution Chromatogr. 17: 415.

Treibs, A. 1935. Ann. Chem. 517: 172.

Vercier, P., and Mounton, M. 1982. Analysis. 10:57.

Waldo, G. S., Carlson, R. M. K., Moldowan, J. M., Peters, K. E., and Penner-Hahn, J. E. 1992. Geochim. Cosmochim. Acta. 55: 801.

Waldo, G. S., Mullins, O. C., Penner-Hahn, J. E., and Cramer, S. P. 1992. Fuel. 71: 53.

Yen, T. F. 1975. In Role of Trace Metals in Petroleum. T. F. Yen (Editor). Ann Arbor Scientific Publishers, Ann Arbor, MI. p. 1.

Chapter 4

Asphaltenes and the Structure of Petroleum

James G. Speight
Western Research Institute
365 North 9th Street
Laramie
Wyoming 82072

1.0 INTRODUCTION

The measurable decline in crude oil quality over the past two decades and the difficulties encountered in recovering and processing heavier, high-sulfur crude oils depend, to a large extent, on the amount of complex, higher boiling constituents present in the feedstock (Speight, 1981, 1984, 1986). Producing and refining such feedstocks is not just a matter of applying know-how derived from work with conventional crude oils but also requires a knowledge of the composition of these more complex feedstocks (Speight, 1992; Speight, 1994).

Petroleum is a mixture of compounds boiling at different temperatures that can be separated into a variety of different generic fractions (Chapters 1 and 2) (Speight, 1991). As the ranges of molecular weight and boiling point of the fractions increase, there is a concentration effect that results in the majority of the heteroatoms and aromatic systems concentrating in the asphaltene and resin fractions with the obvious outcome that these two fractions produce most of the thermal coke that is formed during processing (Chapter 5).

Inasmuch as there is a wide variation in the properties of crude petroleum, the proportions of the different fractions vary with origin (Speight, 1991). Variations in composition notwithstanding, petroleum can be fractionated into several bulk fractions, but the active principle is that petroleum is a continuum and perturbations to the system can cause the separation of a solid phase (Long, 1979, 1981; Speight, 1991, 1994). Indeed, the asphaltene fraction (the predominant cause of sediment formation) is ubiquitous in petroleum and has received considerable attention over the past five decades.

Thus, a key to understanding crude oil behavior during refining operations is to develop an understanding of the behavior of the various constituents under those conditions of temperature and pressure used in the various processes.

Indeed, a knowledge of the chemical makeup and the resulting chemical behavior is crucial. This is especially true for the asphaltenes and resins and the means by which they are converted to liquids and to coke (Chapter 5) and to other sediments (Chapter 8).

However, in a mixture as complex as a residuum or heavy oil, refining chemistry can only be generalized, because it is difficult to analyze not only the products but also the feedstock as well as the intricate and complex nature of the molecules that make up the feedstock. Also, refining is an area that remains largely undefined, insofar as the chemistry and physics are still speculative, especially when asphaltene-containing feedstocks are involved.

The presence of asphaltenes in feedstocks is particularly acute when the circumstances permit their separation (precipitation) from the feedstock in such a manner that the phenomenon of phase separation causes coke formation and catalyst deterioration during processing.

Briefly, the asphaltene fraction is the most refractory (or the highest molecular weight fraction of petroleum (see Section 5.0 in this chapter). A generally accepted definition of asphaltenes is in terms of their solubility in low-boiling liquid hydrocarbons (Chapter 1; see also Section 2.0 in this chapter). The asphaltene fraction, because of its refractory nature, can be refined only with difficulty but it is established that thermal treatment of the asphaltenes will produce approximately 50% volatile materiel and 50% coke (Speight, 1970). Thus, in many vacuum residua, where the asphaltene fraction is a major component, it is the propensity of the asphaltenes to produce high yields of thermal coke that is of concern.

The ultimate goal is to convert the asphaltenes to useful products while preventing process deterioration through coke formation or catalyst inefficiency (Speight, 1991). For both cases, an understanding of the fundamental chemistry of the asphaltene constituents can be rewarding.

The complex nature of the asphaltene fraction can be taken several steps further, when it is acknowledged that the constituents are *multidimensional in space* in contrast to the two-dimensional aspects often represented, for convenience, *on paper* (Speight, 1994). Such knowledge will also offer new insights into the thermal chemistry of asphaltenes (Chapter 5).

In addition to understanding the chemical reactions of the asphaltene and resin constituents, another key aspect of understanding crude oil behavior during refining operations is to develop an understanding of the principal precursors (the asphaltenes and other polar constituents) to coke (Chapter 5) and the means by which these polar constituents form coke and other sediments (Chapter 8). Further understanding might be added through investigations of the relationships between the asphaltenes and the other constituents of petroleum feedstocks. Such investigations will help in developing behavioral traits and afford some degree of predictability of the outcome of refining operations.

This chapter summarizes the data related to asphaltene character and the interrelationships of the constituents of petroleum. This can then be used to explain various behavioral aspects, including the onset of coke formation, the events that lead to sediment formation in cracked and visbroken feedstocks, catalyst deactivation, and related phenomena.

2.0 SEPARATION

Asphaltenes are, by definition, a solubility class that is precipitated from petroleum, heavy oil, and bitumen by the addition of an excess of a liquid hydrocarbon such as *n*-pentane or *n*-heptane, and there are several standard methods for asphaltene precipitation (Table 4-1) (Girdler, 1965; Mitchell and Speight, 1973; Speight et al., 1984; Andersen and Birdi, 1990; Speight, 1991; Speight, 1994) (Chapter 1).

Table 4-1. Standard methods for asphaltene precipitation.

Method	Precipitant	Volume precipitant per g of sample
ASTM D-893	*n*-pentane	10 ml
ASTM D-2006	*n*-pentane	50 ml
ASTM D-2007	*n*-pentane	10 ml
ASTM D-3279	*n*-heptane	100 ml
ASTM D-4124	*n*-heptane	100 ml
IP 143	*n*-heptane	30 ml
Syncrude method	*n*-pentane	20 ml
UOP 99-82	*n*-pentane	not known

ASTM Annual Book of Standards, 1980-1994.
Cooley et al., 1988.
Syncrude Research, 1979.

As mentioned earlier, the nomenclature of the petroleum constituents is an *operational aid* and is not usually based on chemical or structural features. Thus, there is no one parameter that is operational in the separation of asphaltenes. The relevant parameters for asphaltene separation are physical and chemical in nature and include:

1. Polarity (the presence of functional groups derived from the presence of heteroatoms in the asphaltenes);
2. Aromaticity (the presence of polynuclear aromatic systems in the asphaltenes);
3. Molecular weight (molecular size);
4. Three-dimensional structure (the micelle) of the asphaltene constituents as they exist in relationship with the other constituents of crude oil;
5. Solvent power of the precipitating/extracting liquid used for the separation (Mitchell and Speight, 1973);
6. Time required to allow the precipitating/extracting liquid to penetrate the micelle, which is dependent upon the ability of the hydrocarbon liquid to penetrate the micelle, indicating that the process is diffusion controlled (Speight et al., 1984);
7. Ratio of the precipitating/extracting liquid to crude oil, which dictates the yield and character of the asphaltene product (Mitchell and Speight, 1973; Speight et al., 1984);
8. Temperature, which may reduce the induction period that is a requirement of diffusion-controlled processes (Mitchell and Speight, 1973); and
9. Pressure, employed in several refinery processes as a means of maintaining the low-boiling liquid hydrocarbon in the liquid phase.

Other parameters may be defined as subsets of those enumerated above, and it is worthy of note that, in order to remove entrained resin material, reprecipitation of the asphaltenes from benzene or toluene is often necessary (Speight et al., 1984; Ali et al., 1985).

The focal point of the asphaltene separation method is the isolation of the asphaltene constituents as a discrete fraction. However, a question that often arises is related to whether or not the asphaltenes self-associate in the crude oil or whether the asphaltene constituents interact directly with the other crude oil constituents. In addition, there is also a question related to precipitation of the asphaltenes or extraction of the constituents from the micelle that are responsible for the dispersion of the asphaltenes in the oil.

The former case, i.e., precipitation, is assumed to be the means by which asphaltenes are separated from light petroleum, whereas extraction is the means by which the asphaltenes are separated from the heavier crude oils. For clarification, the lower molecular weight constituents in heavy feedstocks are presumed to be extracted by the separating hydrocarbon. If the micelle concept is true, extraction of the lower molecular weight constituents from the micelle should occur whatever the specific gravity or viscosity of the feedstock. The extraction concept is most likely.

The precipitation of asphaltenes can be ascribed to changes in crude oil composition caused by the addition of lower boiling components that alter the complex equilibrium keeping the asphaltenes in solution or in a *peptized state*.

In the case of addition of low-boiling liquid hydrocarbons to the feedstock, the hydrocarbon causes a change in the solubility parameter of the oil medium that, in turn, changes the tolerance of the medium for the complex micelle structure. As this occurs, the lower molecular weight and less polar constituents of the micelle are extracted into the liquid, leaving the asphaltene constituents without any surrounding (dispersing) sheath. Separation then ensues. If the process is thermal (i.e., in excess of the thermal decomposition temperature), a different sequence of events ensues leading to the formation of an insoluble phase and ultimately to coke (Chapter 5).

3.0 COMPOSITION

The asphaltene fraction is composed of many different types of molecular species (Speight, 1991). Therefore, no one description can be used for asphaltenes, and definitions of asphaltenes are based on descriptors such as high molecular weight constituents, high-polarity constituents, and refractory constituents. On a molecular level, the descriptors are even more numerous and more complex. Of course, all of these descriptors are relative to the other constituents of petroleum.

Analytical data from various techniques and methods can produce only *average values*; such data do show that the asphaltene fraction contains the most aromatic and most polar species originally present in the crude oil. However, caution must be used in the application of average data to structural studies where the formula for an average molecule might be derived. An *average molecule* that might be derived from the data may be very far from the truth and may explain asphaltene behavior only with difficulty and a lot of imagination!

3.1 Ultimate Analysis (Elemental Composition)

The elemental compositions of asphaltenes vary over a narrow range corresponding to H/C ratios of 1.15 ± 0.05%, although values outside this range are often found (Speight, 1991). Notable variations do occur in the proportions of the heteroelements, in particular in the proportions of oxygen and sulfur. The nitrogen content of the asphaltenes has a somewhat lesser degree of variation. This is, perhaps, not surprising, since exposing asphaltenes to atmospheric oxygen can substantially alter the oxygen content and exposing a crude oil to elemental sulfur, or even to sulfur-containing minerals, can result in sulfur uptake.

3.2 Fractional Composition

Asphaltenes can be fractionated by a variety of techniques (Bestougeff and Darmois, 1947, 1948; Altgelt, 1965; Bestougeff and Mouton, 1977; Francisco and Speight, 1984). Of specific interest is the observation that when asphaltenes are fractionated on the basis of aromaticity and polarity, it appears that the more aromatic species contain higher amounts of nitrogen (Speight, 1984), suggesting that the nitrogen species are located predominantly in aromatic systems.

The solvent-based fractionation of the asphaltene component of petroleum showed that it is possible to obtain asphaltene fractions characterized by different degrees of aromaticity or heteroatom content by using benzene/pentane or benzene/methanol mixtures in variable ratios (Speight, 1979). The use of mixtures of a polar and a nonpolar solvent in order to fractionate an asphaltene sample, however, will tend to direct the fractionation by introducing polar forces and hydrogen bonding, as well as dispersion forces, as factors determining which components of the asphaltene sample are soluble in the mixture.

In contrast, the exclusive use of saturated hydrocarbons, whose solvent power derives entirely from the dispersion forces, should in principle permit the fractionation to be carried out with variation of only a single component of the solubility parameter. It is known that the quantity of asphaltenes precipitated by linear paraffins decreases with increasing carbon number until a limiting value is reached above *n*-heptane (n-C_7H_{16}) to *n*-nonane (n-C_9H_{20}) (Girdler, 1965; Mitchell and Speight, 1973).

The cycloparaffins have better solvating power toward the asphaltenes and often fail to induce asphaltene separation from crude oil, but it is possible to fractionate asphaltenes using mixtures of solvents, such as cyclohexane and *n*-alkanes, and any differences in the precipitation threshold underscores the point that the solubility properties of the asphaltene fractions of different crude oils can vary markedly (Acevedo et al., 1995).

It appears, therefore, that the tendency of asphaltenes to precipitate from a given crude oil because of a variation in the solubility parameter of the precipitating solvent mixture is more closely related to the aromaticity and polarity of the asphaltenes than to their dimensions. However, it is possible to conclude (Acevedo et al., 1995) that molecular weight, aromaticity, and polarity combine to determine asphaltene solubility in hydrocarbon media.

The fractionation of asphaltenes into a variety of functional (and polar) types (Figure 4-1) (Francisco and Speight, 1984) has confirmed the complexity of the asphaltene fraction. High-performance liquid chromatography (HPLC) has also been used to illustrate the diversity of the structural and functional types in asphaltenes to the extent that different high-performance liquid chromatographic profiles can be expected for different asphaltenes (Speight, 1986, 1991).

In addition, the molecular weight/polarity concept can be used to indicate the relative nature of carbenes and carboids (thermal decomposition products of asphaltenes), which are lower molecular weight and more polar than their precursors (Figure 4-2) (Chapter 1). However, it should not be construed that this two-dimensional diagram is truly representative of the nature of asphaltenes and carbenes and carboids. A three-dimensional diagram involving functionality or molecular weight might be more appropriate for showing the chemical and physical differences between the various asphaltenes, carbenes, and carboids. However, the concept does emphasize the lower molecular weight and increased polarity of the various constituents of asphaltenes.

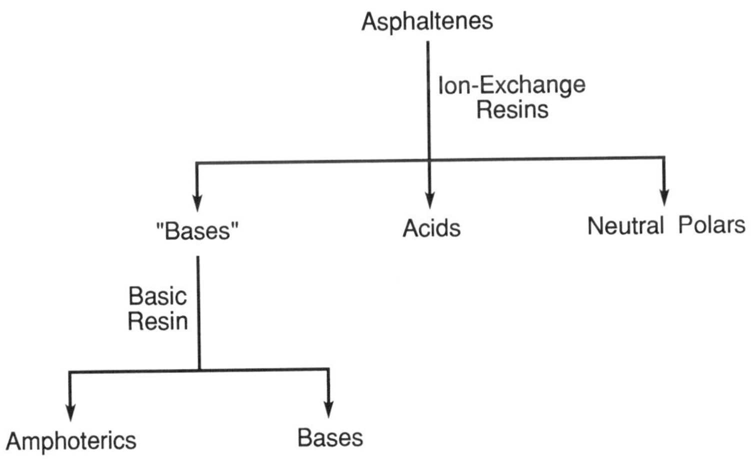

Figure 4-1. The fractionation of asphaltenes into a variety of functional (and polar) types.

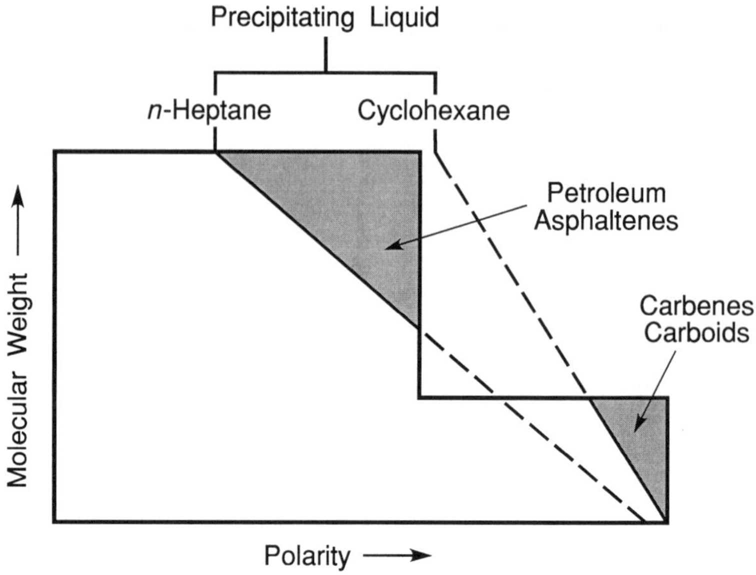

Figure 4-2. Representation of the differences in polarity and molecular weight between asphaltenes, carbenes, and carboids.

In summary, fractionation data lend support to, and reinforce, the concept that asphaltenes are complex mixtures of molecular sizes and various functional types (Long, 1979, 1981) and that carbenes and carboids can also fit into this concept (Speight, 1994).

4.0 STRUCTURAL STUDIES

The molecular nature of the nonvolatile fractions of petroleum has been the subject of numerous investigations (Speight, 1972; Yen, 1972, 1974; Speight, 1991, 1994). The data from these studies indicated that asphaltenes, viewed structurally, contain condensed polynuclear aromatic ring systems bearing alkyl side chains.

It was generally believed that as the boiling point of the crude oil fraction increased, the size of (i.e., the number of rings in) the polynuclear aromatic systems increased, although (on the basis of fractionation studies) it would be more appropriate to place the asphaltenes in the less well-defined region marked *polar aromatics*. This is generally true, to a point. But it is likely that the size of the condensed ring system is more in keeping with the skeletal structures of natural product ring systems, where the number of condensed rings rarely exceeds six. In fact, it is more likely that within the resin and asphaltenes fraction contributions to the higher boiling point involve the inclusion of heteroatom systems, longer alkyl side chains, and naphthenic systems. This would be more consistent with the concept that petroleum is a continuum of chemical types (Speight, 1994).

Thus, with the concept of larger polynuclear aromatic systems in mind, efforts were made to describe the total structures of asphaltenes in accordance with magnetic resonance data and results of other spectroscopic and analytical techniques. The macrostructure of the asphaltenes was believed to involve graphite-like stacks (see Speight, 1994 and references cited therein).

The focus of studies related to the structure of asphaltenes has been on the carbon skeleton. However, the presence of functional groups contributes strongly to the physical-chemical properties of the asphaltenes (Speight, 1991; Altgelt and Boduszynski, 1994).

Nitrogen occurs in asphaltenes in various heterocyclic species (Chapter 3) (Speight, 1991, 1994; Mullins, 1995). The more conventional types of nitrogen, i.e., primary, secondary, and tertiary amines, have not been established as being present in petroleum asphaltenes. The nitrogen has also been defined in terms of basic and nonbasic types: pyridine derivatives (basic) and indoles, carbazoles, and amides as well as porphyrinic nitrogen (neutral) (McKay et al., 1976; Moschopedis et al., 1981; Jacobsen and Gray, 1987). The application of x-ray absorption near-edge structure (XANES) spectroscopy to the study of asphaltenes has led to the conclusion that a large portion of the nitrogen present is aromatic, in pyrrolic rather than pyridinic form (Mitra-Kirtley et al., 1993).

Oxygen has been identified in carboxylic, phenolic, and ketonic locations (Nicksic and Jeffries-Harris, 1968; Petersen et al., 1974; Ritchie et al., 1979; Speight and Moschopedis, 1981; Moschopedis et al., 1981; Rose and Francisco, 1987).

Sulfur, the predominant heteroatom in crude oil, occurs principally in thiophenic structures (i.e., thiophene, benzothiophene dibenzothiophene, and naphthobenzothiophene), as well as in aliphatic structures (alkyl-alkyl sulfides, alkyl-aryl sulfides, and aryl-aryl sulfides), thiane and thiolane systems, and oxidized forms (Drushel, 1970; Yen, 1974; Speight and Pancirov, 1984; Rose and Francisco, 1988; Keleman et al., 1990) (see also Chapter 3). Disulfides may also be present in petroleum but have not been detected to the same extent as the other sulfur systems; they might be expected to dissociate under maturation conditions or during the exposure to high temperatures (340°C, 640°F) to which crude oil is subjected during distillation. More highly condensed thiophene types may also exist but are precluded from identification by low volatility. Investigations of the size distribution of the sulfur species indicate an increase in the size of the sulfur species from aromatics to resins to asphaltenes (Speight, 1991, 1994).

Metals (i.e., nickel, vanadium, and iron) also occur in petroleum (Baker, 1969; Yen, 1975) and are concentrated in the heaviest part of the crude oil, but whether these are an integral part of the asphaltene structure is not known. A significant part of this organometallic component is porphyrinic in nature, although many other structures, not yet better identified, are also present (Chapter 3). Some of the porphyrin systems can be isolated as a separate stream from petroleum (Branthaver, 1990) (Chapter 3).

5.0 MOLECULAR WEIGHT

The molecular weights of asphaltenes span a wide range from a few hundred to several million, leading to the speculation about asphaltene self-association (Speight et al., 1995).

The tendency of the asphaltenes to form aggregates in hydrocarbon solution is one of their most characteristic features and complicates the determination of asphaltene molecular weight. The average molecular weights measured by means of vapor pressure osmometry (VPO) or size exclusion chromatography (SEC) are significantly influenced by the conditions of the analysis (temperature, asphaltene concentration, solvent polarity) (Altgelt, 1968; Moschopedis et al., 1976; Speight and Moschopedis, 1980; Speight, 1981; Speight et al., 1985; Acevedo et al., 1992). For this reason, molecular weights are reported in the literature as relative values, and these values may be quite different from the molecular weights of unassociated molecules.

The influence of the asphaltene concentration on the values of molecular weight measured is significant and it is important to use as high a dilution as possible in order to measure a molecular weight that at least approaches that of the unassociated asphaltene molecules, remembering that the asphaltene fraction is a complex mixture of different species (Chapter 1) and that any molecular weight of unassociated species will be an average molecular weight of different chemical species.

By means of small-angle neutron scattering (SANS) studies, the existence of aggregates in oil has been demonstrated, and the physical dimensions and shape of the aggregates are functions of the solvent used (Ravey et al., 1988; Overfield et al. 1989). In addition, surface tension measurements have been used to study the self-association of asphaltenes in pyridine and nitrobenzene (Sheu et al., 1992).

A discontinuous transition in the surface tension as a function of asphaltene concentration was interpreted as the critical asphaltene concentration above which self-association occurs.

In summary, asphaltene molecular weights are variable (Yen, 1974; Speight et al., 1985), there being a tendency to associate even in dilute solution in nonpolar solvents. However, data obtained using highly polar solvents indicate that the molecular weights, in solvents that prevent association, usually fall in the range 2000 ± 500.

6.0 MOLECULAR MODELS

Asphaltenes are macromolecules insofar as the molecular weights are generally accepted to be at least several hundred and more often approaching 2000. In addition, asphaltenes exist as complex mixtures of components that are separated from petroleum by treatment with a low-boiling liquid hydrocarbon (Mitchell and Speight, 1973; Speight et al., 1984)

A number of molecular models have been proposed for asphaltenes and are described in detail elsewhere (Speight, 1994), and repetition is not warranted here. It is sufficient to state that any molecular models derived for asphaltenes must be in keeping with behavioral characteristics. Efforts were made to describe the asphaltene fraction in terms of a single, representative asphaltene molecule or molecules (Speight, 1994) incorporating, in the correct proportions, all of the chemical and elemental constituents known to be present in a given asphaltenic matrix.

Although this approach gives an idea of the structural complexity of the compounds making up the asphaltene component, it obscures the highly differentiated chemical nature of the molecules in this petroleum fraction (Long, 1979, 1981). The idealized molecular structures defined by this method, however, are of some utility for inferring the chemistry and physical-chemical properties of the asphaltenes (Speight, 1992, 1994).

Briefly, there are essentially three types of models:

1. Those that simply help to visualize the three-dimensional architecture and stereochemistry but are not to scale.
2. Framework-type models that indicate correct bond distances and bond angles and can be used to measure distances between nonbonded atoms in molecules but do not show the atoms as such.
3. So-called space-filling models that provide a fairly realistic three-dimensional representation of what the molecule actually looks like.

Attempts to model asphaltenes have usually employed pictographic representations of the first kind of models. Minor efforts have been made to represent asphaltenes using models of types 2 and 3.

A serious shortcoming, common to virtually all molecular models, is that they have fixed bond angles and that rotation about single bonds is excessively facile, especially in the first two types of models. In contrast, in the real situation there are relatively easily definable bond angles and substantial barriers to rotation about single bonds. Moreover, if it is desired to measure intramolecular and/or intermolecular bond distances, the model must first be fixed in its actual conformation. For molecules having a number of single bonds, this may be quite inconvenient (even though mechanical devices to stop bond rotation are available) and it may be difficult to set the actual torsion angles with any kind of precision.

As a result of all these difficulties, modeling of the macromolecules in the asphaltene fraction has proved to be speculative, although some success has been achieved in attempting to understand the nature of the constituents and their behavior in thermal processes (Speight, 1994). Success notwithstanding, the chemical dynamics are still speculative because of the speculative nature of the model!

The properties of macromolecules depend on both the chemical structure and physical structure. The term *primary structure* describes the chemical sequence of atoms in a macromolecule. The ordering of the atoms in space relative to each other is referred to as the *secondary structure*, and the three-dimensional structure of a molecule is called the *tertiary structure*.

By analogy, the *primary structure* of asphaltenes is the two-dimensional structure derived by a variety of analytical techniques and is often presented on paper as an *average structure*. The *secondary and tertiary structure*, and perhaps an often ignored but extremely important aspect of asphaltene chemistry and physics, is the micelle structure, which represents the means by which asphaltenes exist in crude oil.

7.0 THE STRUCTURE OF PETROLEUM

The pronounced tendency of asphaltenes to form aggregates in hydrocarbon solution is one of their most characteristic traits. Thus, while a number of experimental methods indicate that isolated asphaltene monomers have an *average* molecular weight on the order of 1500 to 2500 (vapor pressure osmometry in polar solvents), numerous methods show that asphaltenes associate spontaneously in most hydrocarbon media, resulting in observed molecular weights of 10,000 or more. Thus, it is not surprising that this tendency for asphaltenes to aggregate in hydrocarbon solvents has been applied to the existence of asphaltenes in crude oil. The debate is about whether this line of thinking is correct!

An early hypothesis of the physical structure of petroleum (Pfeiffer and Saal, 1940) indicated that asphaltenes are the centers of micelles formed by adsorption, or even by absorption of part of the maltenes, that is, resin material, on the surfaces or in the interiors of the asphaltene particles. On the basis of this hypothetical picture, resins or maltenes are necessary in order to suspend the asphaltene core, and, thus, the structure of the micelle is an asphaltene-resin complex rather than an asphaltene-asphaltene micelle.

A later model of asphaltene micelle structure was proposed in which the asphaltenes, having a strong propensity for self-association, formed self-associated cores as the central unit of the micelle (Dickie and Yen, 1972). Within a unit, the polynuclear aromatics of the asphaltene molecules stack to form local discotic structures.

However, the vast differences in molecular structure among individual asphaltene molecules prevent these agglomerates from forming a discotic phase, which requires the monomers to have identical molecular structures. Instead, the asphaltenes randomly stack to form globular micelles as proposed in this model. It was also proposed that the elementary micelle units can further cluster into larger particles similar to flocs.

Acceptance of these models and the concept of a micelle is easier to understand if the constituents of the resin and asphaltene fractions are viewed as two adjacent portions of a single broad compositional continuum that contains the polar, aromatic, and higher molecular weight components of petroleum (Speight, 1991). This allows the further acceptance of structural similarities in the asphaltenes and resins that facilitate formation of the micelle(s) (Speight, 1991, 1994).

The means by which asphaltenes associate in solution and in crude oil has been the subject of many investigations (Speight, 1991). Much of the work related to asphaltene association has arisen from molecular weight measurements from which deductions have been made about the asphaltene micelle, but there is still considerable conjecture about the true nature of the micelle. Specifically, there is some debate about whether the micelle in petroleum is composed of homogeneous material, insofar as it is composed only of asphaltene molecules, or is composed of asphaltene and resin molecules (Pfeiffer and Saal, 1940; Dickie and Yen, 1967; Koots and Speight, 1975; Speight, 1991).

If the composition and properties of the precipitated asphaltenes reflect those of the micelles in solution, the latter should be considered as mixed micelles. This is a reasonable assumption in view of the large quantities of soluble resins found in the precipitated solid. It should be noted that if mole fractions are used as a composition measure, the one for resins in the precipitates will be greater than that for asphaltenes.

Considering the micelles to be composed of asphaltene species that are *peptized* by resins and asphaltenes is a more reasonable approach. An important corollary of petroleum composition is that the mole fraction of resins is always larger than that of asphaltenes and hence the micelles are expected to be richer in resins (Figure 4-3). The center or core of the micelle would be occupied by highly polar asphaltene molecules that are surrounded by other more soluble asphaltene molecules, which would be placed between the center and the periphery. The inclusion of other asphaltene species is not seen as a graphite-type stack, as has been proposed previously. It is more likely an association of convenience that facilitates the association of the central asphaltene with resin species. In many cases, more than one asphaltene molecule per micelle is considered unlikely.

However, in such a model, the substances with higher molecular weights and with the most pronounced aromatic nature are situated closest to the nucleus and are surrounded by lighter constituents of less aromatic nature. The transition of the intermicellar (dispersed or oil) phase is gradual and almost continuous.

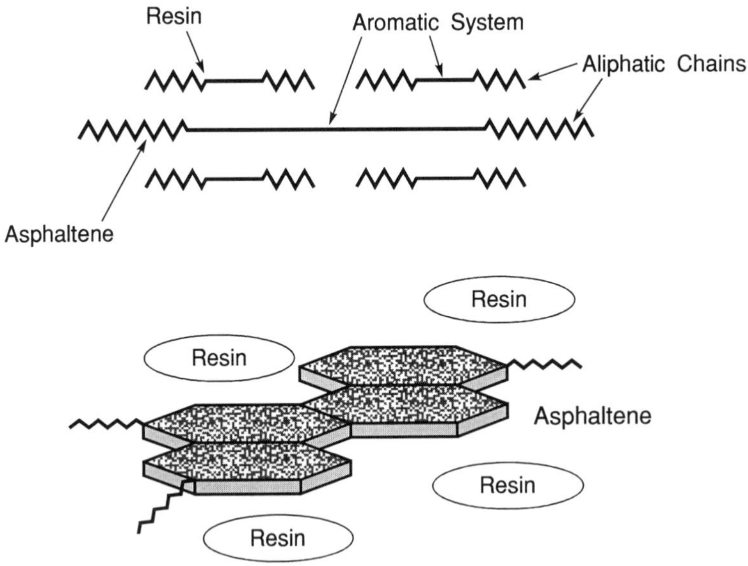

Figure 4-3. Representation of an asphaltene-resin micelle.

Since asphaltenes are incompatible with the oil fraction (Swanson, 1942; Koots and Speight, 1975), asphaltene dispersion is attributable mainly to the resins (polar aromatics).

Empirical observations indicate that the resins play an important role in stabilizing asphaltenes in crude oil (Koots and Speight, 1975 and references cited therein), and under unfavorable solvent conditions the asphaltene species are prone to further aggregation into clusters that are unstable and precipitate from the crude oil (Mitchell and Speight, 1973; Koots and Speight, 1975; Yen, 1990).

$$\text{asphaltene molecule} \leftrightarrow \text{micelle} \leftrightarrow \text{micelle clusters/flocs} \qquad (1)$$

This model requires that the asphaltene micelles are composed of an insoluble molecular core that associates with the resins, thereby providing steric stabilization against flocculation and precipitation. Thus, the phase separation of asphaltenes upon the addition of nonpolar solvents to the crude oil can be rationalized in terms of reduction of the solubility parameter or polarity of the hydrocarbon medium (Speight, 1991; Wiehe 1992, 1993). This results in solubilization of the resins, leading to dissociation of the resin-asphaltene complexes, destabilization of the soon-to-be-unattended asphaltenes, and flocculation, precipitation, and phase separation.

An alternative explanation for the stabilizing influence of the resins, and for the stability of asphaltenes in hydrocarbon media in general, is that individual asphaltene species and micelles are in thermodynamic equilibrium (Cimino et al., 1995). Thus, the degree of association is determined by the relative energies of solvation of the monomers and the micelles and the entropy and enthalpy changes associated with the self-association of the asphaltene monomers.

According to this model, the stabilization of the asphaltene micelles by the resin fraction derives from the contribution of the latter to the solvent power (polarity and aromaticity) of the medium. Destabilization of the asphaltenes, leading to phase separation, occurs when the solvating power of the medium toward the asphaltene monomers and micelles is reduced to the point at which they are no longer fully soluble.

The size of the asphaltene micelles varies as a function of the temperature, the asphaltene concentration, and the identity of the solvent, thereby establishing that the self-association of the asphaltenes is reversible and that the sizes in the solution state result from a true thermodynamic equilibrium between primary particles and reversible aggregates, the equilibrium constant depending on some property of the solvent. But, again, the relationship of the behavior of asphaltenes in hydrocarbon media to their behavior in crude oil is still speculative.

Thus, the means by which asphaltenes exist in crude oil is assumed to involve asphaltene-resin interactions leading to the formation of micelles (Koots and Speight, 1975), but there are still many unknowns related to the existence of asphaltenes in crude oil.

On this basis, the stability of petroleum can be represented by a three-phase system in which the asphaltenes, the aromatics (including the resins), and the saturates are delicately balanced (Speight, 1991). Various factors, such as oxidation, can have an adverse effect on the system, leading to instability or incompatibility as a result of changing the polarity, and bonding arrangements, of the species in crude oil.

8.0 REFERENCES

Acevedo, S., Escobar, G., Gutierrez, L. B., and D'Aquino, J. 1992. Fuel. 71: 1077.

Acevedo, S., Ranaudo, M. A., Escobar, G., Gutierrez, L. B., and Gutierrez, X. 1995. In Asphaltenes: Fundamentals and Applications. E. Y. Sheu and O. C. Mullins (Editors). Plenum Press, New York. Chapter IV.

Ali, L. H., Al-Ghannan, A., and Speight, J. G. 1985. Iraqi J. Sci. 26: 41.

Altgelt, K. H. 1968. Preprints Div. Petrol. Chem. Am. Chem. Soc. 13(3): 37.

Altgelt, K. H., and Boduszynski, M. M. 1994. The Composition and Analysis of Heavy Petroleum Fractions. Marcel Dekker Inc., New York.

Andersen, S. I., and Birdi, K. S. 1990. Fuel Sci. Technol. Int. 8: 593.

Baker, E. W. 1969. In Organic Geochemistry. G. Eglinton and M. T. J. Murphy (Editors). Springer-Verlag, New York.

Bestougeff, M. A., and Darmois, R. 1947. Comptes Rend. 224: 1365.

Bestougeff, M. A., and Darmois, R. 1948. Comptes Rend. 227: 129.

Bestougeff, M. A., and Mouton, Y. 1977. Bull. Liason Lab. Pont. Chaus. Special Volume. p. 79.

Branthaver, J. F. 1990. In Fuel Science and Technology Handbook. J.G. Speight (Editor). Marcel Dekker Inc., New York.

Cimino, R., Correrra, S., Del Bianco, A., and Lockhart, T. P. 1995. In Asphaltenes: Fundamentals and Applications. E. Y. Sheu and O. C. Mullins (Editors). Plenum Press, New York. Chapter III.

Cooley, J., Montgomery, D. S., Pickard, A. L., Hoffman, E. R., Lynch, T., Litka, S. E., and Thomas, K. P. 1988. In A Review of Analytical Methods for Bitumens and Heavy Oils. D. Wallace (Editor). AOSTRA Technical Publication Series No. 5. Alberta Oil Sands and Research Authority, Edmnton, Alberta, Canada. p. 100.

Dickie, J. P., and Yen, T. F. 1972. Anal. Chem. 39: 1487.

Francisco, M. A., and Speight, J. G. 1984. Preprints Div. Fuel. Chem. Am. Chem. Soc. 29(1): 36.

Girdler, R. B. 1965. Proc. Assoc. Asphalt Paving Technol. 34: 45.

Jacobsen, J. M., and Gray, M. R. 1987. Fuel. 66: 749.

Keleman, S. R., George, G. N., and Gorbaty, M. L. 1990. Fuel. 69: 939.

Koots, J. A., and Speight, J. G. 1975. Fuel. 54: 179.

Long, R. B. 1979. Preprints Div. Petrol. Chem. Am. Chem. Soc. 24(4): 891.

Long, R. B. 1981. In The Chemistry of Asphaltenes. J. W. Bunger and N. Li (Editors). Advances in Chemistry Series No. 195. American Chemical Society, Washington, DC

McKay, J. F., Weber, J. H., and Latham, D. R. 1976. Anal. Chem. 48: 891.

Mitchell, D. L., and Speight, J. G. 1973. Fuel. 52: 149.

Mitra-Kirtley, S., Mullins, O. C., van Elp, J., George, S. J., Chen, J., and Cramer, S. P. 1993. J. Am. Chem. Soc. 115: 252.

Moschopedis, S. E., Fryer, J. F., and Speight, J. G. 1976. Fuel. 55: 227.

Moschopedis, S. E., Hawkins, R. W., and Speight, J. G. 1981. Fuel. 60: 397.

Mullins, O. C. 1995. In Asphaltenes: Fundamentals and Applications. E. Y. Sheu and O. C. Mullins (Editors). Plenum Press, New York. Chapter II.

Nicksic, S. W., and Jeffries-Harris, M. J. 1968. J. Inst. Pet. London. 54: 107.

Overfield, R. E., Sheu, E. Y., Sinha, S. K., and Liang, K. S. 1989. Fuel Sci. Technol. Int. 7: 611.

Petersen, J. C., Barbour, F. A., and Dorrence, S. M. 1974. Proc. Assoc. Asphalt Paving Technol. 43: 162.

Pfeiffer, J. P., and Saal, R. N. 1940. Phys. Chem. 44: 139.

Ravey, J. C., Decouret, G., and Espinat, D. 1988. Fuel. 67: 1560.

Ritchie, R. G. S., Roche, R. S., and Steedman, W. 1979. Fuel. 58: 523.

Rose, K. D., and Francisco, M. A. 1987. Energy and Fuels. 1: 233.

Rose, K. D., and Francisco, M. A. 1988. J. Am. Chem. Soc. 110: 337.

Sheu, E. Y., DeTar, M. M., Storm, D. A., and DeCanio, S. J. 1992. Fuel. 71: 299.

Speight, J. G. 1970. Fuel. 49: 134.

Speight, J. G. 1972. Appl. Spectrosc. Rev. 5: 211.

Speight, J. G. 1979. Studies on Bitumen Fractionation. Information Series No. 84. Alberta Research Council, Edmonton, Alberta, Canada.

Speight, J. G. 1981. Preprints Div. Petrol. Chem. Am. Chem. Soc. 26: 825.

Speight, J. G. 1984. In Characterization of Heavy Crude Oils and Petroleum Residues. B. Tissot (Editor). Editions Technip, Paris.

Speight, J. G. 1986. Preprints Div. Petrol. Chem. Am. Chem. Soc. 31: 818.

Speight, J. G. 1987. Preprints Am. Chem. Soc. Div. Petrol. Chem. 32(2): 413.

Speight, J. G. 1991. The Chemistry and Technology of Petroleum. 2nd Edition. Marcel Dekker Inc., New York.

Marcel Dekker Inc., New York.
Speight, J. G. 1992. Proceedings. 4th International Conference on the Stability and Handling of Liquid Fuels. U.S. Department of Energy (DOE/CONF-911102). p. 169.
Speight, J. G. 1994. The Chemistry and Technology of Coal. 2nd Edition. Marcel Dekker Inc., New York.
Speight, J. G., and Moschopedis, S. E. 1980. Fuel. 59: 440.
Speight, J. G., and Moschopedis, S. E. 1981. Preprints Div. Petrol. Chem. Am. Chem. Soc. 26(4): 907.
Speight, J. G., and Pancirov, R. J. 1984. Liq. Fuels Technol. 2: 287.
Speight, J. G., Long, R. B., and Trowbridge, T. D. 1984. Preprints. Div. Fuel Chem. Am. Chem. Soc. 27: 268.
Speight, J. G., Wernick, D. L., Gould, K. A., Overfield, R. E., Rao, B. M. L., and Savage, D. W. 1985. Rev. Inst. Fr. Petrol. 40: 51.
Swanson, J. M. 1942. J. Phys. Chem. 46: 141.
Syncrude Research. 1979. Syncrude Analytical Methods for Oil Sand and Bitumen Processing. Alberta Oil Sands Technology and Research Authority, Edmonton, Alberta Canada. p. 121.
Wiehe, I. A. 1992. Ind. Eng. Chem. Res. 31: 530.
Wiehe, I. A. 1993. Ind. Eng. Chem. Res. 32: 2447.
Yen, T. F. 1972. Preprints Div. Pet. Chem. Am. Chem. Soc. 17(4): F102.
Yen, T. F. 1974. Energy Sources. 1: 447.
Yen, T. F. 1975. The Role of Trace Metals in Petroleum. Ann Arbor Science Publishers, Inc., Ann Arbor, Michigan.
Yen, T. F. 1990. Preprints Div. Petrol. Chem. Am. Chem. Soc. 35: 314.

Chapter 5

Thermal Chemistry of Petroleum Constituents

James G. Speight
Western Research Institute
365 North 9th Street
Laramie
Wyoming 82072

1.0 INTRODUCTION

Crude oil is rarely used in its raw form but must instead be processed into its various products, generally as a means of forming products with a hydrogen content different from that of the original feedstock (Figure 5-1).

Aside from contaminant minerals such as sulfur and small amounts of trace metals, which are removed during refining, petroleum is composed of hydrocarbons, essentially varying combinations of carbon and hydrogen atoms.

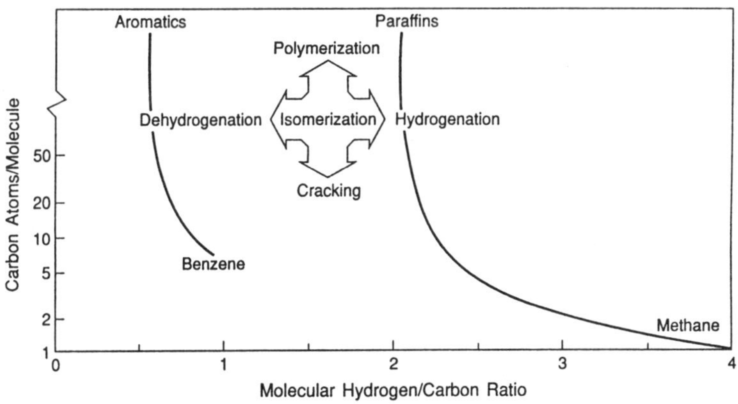

Figure 5-1. Illustration of the chemical changes that occur during petroleum refining.

Any hydrocarbon can be converted into any other hydrocarbon by the appropriate applications of energy, chemistry, and technology. The smaller the molecule and the lower the ratio of carbon to hydrogen, the lighter the hydrocarbon, the lower the evaporation temperature, and, usually, the more valuable the product.

A refinery is a complex network of integrated unit processes for the purpose of producing a variety of products from crude oil (Figure 5-2). Refined products establish the order in which the individual refining units will be introduced, and the choice from among several types of units and the size of these units is dependent upon economic factors. The trade-off among product types, quantity, and quality influences the choice of one kind of process unit over another.

Each refinery has its own range of preferred crude oil feedstock from which a desired distribution of products is obtained. The crude oil is usually identified by its source country, underground reservoir, or some distinguishing physical or chemical property.

Petroleum is a mixture of different hydrocarbons, among which occur various heteroatomic compounds containing nitrogen, oxygen, sulfur, and metals. The two tasks of a refinery are to separate the constituents into usable products and to convert the less desirable constituents into more valuable products.

Distillation is the primary method used to refine petroleum. When the heated crude oil is fed into the lower part of a tower, the lighter oil portions, or fractions, vaporize. Losing temperature as they rise, they condense into liquids, which flow downward into the higher temperatures and are again vaporized. This process continues until the various fractions have achieved the appropriate degrees of purity.

The lighter (lower boiling) constituents, such as fuel gas, propane, butane, gasoline, and kerosene, are taken from the top of the tower and the heavier (higher boiling) constituents, such as fuel oil and gas oil, are taken from the lower parts of the tower.

The characterization of the distillate fractions has been well documented over the past several decades (Speight, 1991), but the character of the higher boiling, even the nonvolatile, constituents has been largely unknown and speculation (often justified) has been applied in the *molecular design* studies (Chapter 4). In this sense, molecular design refers to the estimation, or design, of molecular structures through the application of a variety of chemical and physical techniques. The result is a designed molecule that is believed to represent the structural types found in particular high-boiling nonvolatile fractions (Speight, 1991, 1994). Such studies may be helpful, but also may result in inaccurate conclusions and molecular design data.

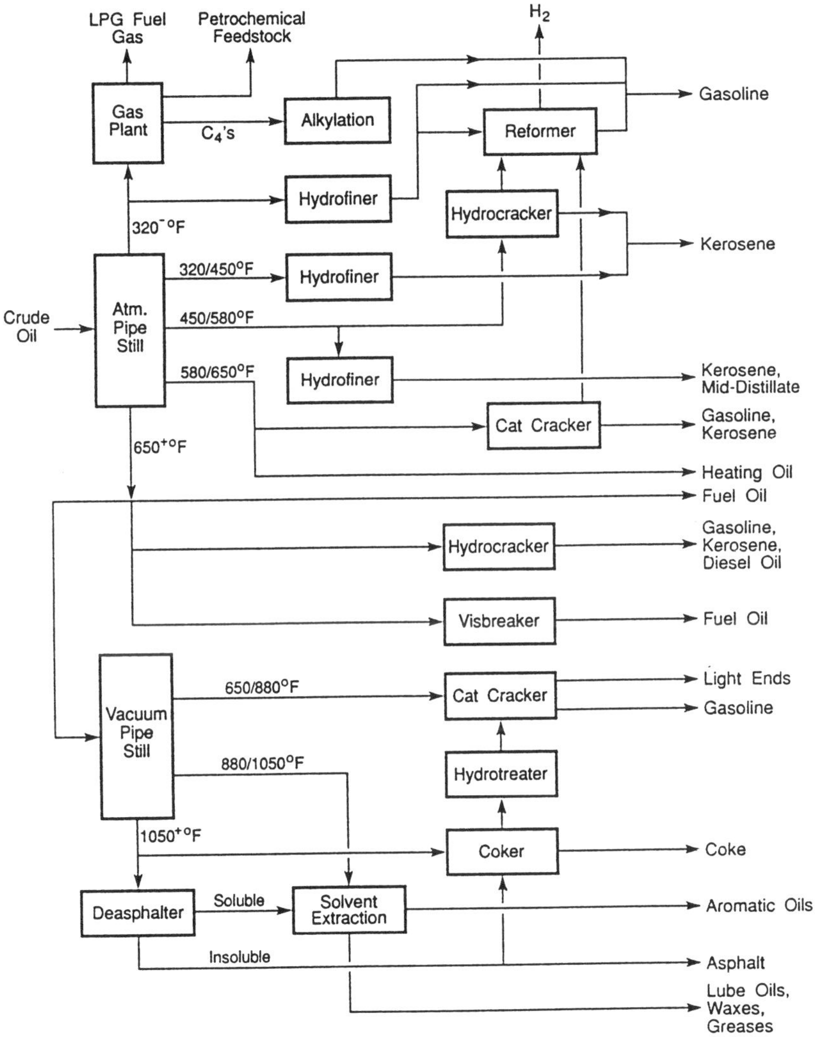

Figure 5-2. Schematic representation of a petroleum refinery.

Constituents of high molecular that are unsuitable for use in the natural state are refined once again through various *conversion* (*thermal decomposition*) processes, generally referred to as *cracking*.

Through the application of vacuum, heat, and catalysts, the higher molecular weight constituents are converted to lower molecular weight products. Thermal cracking, for instance, uses heat and pressure; catalytic cracking employs a finely powdered catalyst; and hydrocracking involves the addition of hydrogen in the presence of a catalyst to produce compounds with higher hydrogen-to-carbon atomic ratios, such as gasoline.

In the early days of petroleum refining, the feedstock was evaluated by means of the American Petroleum Institute (API) gravity, which is a measure of the density that is related to the specific gravity:

$$\text{API} = (141.5/\text{sp. gr.}) - 131.5$$

where sp gr is the specific gravity, or the ratio of the weight of a given volume of oil to the weight of the same volume of water at a standard temperature, usually 15.6°C (60°F).

A crude oil with a density the same as that of water, or with a specific gravity of 1.0, would then be a 10° API oil. Oils with higher than 10° API gravity are lighter than water. Because lighter crude oil fractions are usually more valuable, a crude oil with a higher API gravity will bring a premium price in the marketplace. However, because of the relative scarcity of the lighter crude oils, heavier crude oils are becoming more common refinery feedstocks. For convenience, heavy crude oil is defined as oil with an API gravity of less than 20.

Another characterization factor that was introduced for use as an index of the chemical character of a crude oil or its fractions is the Watson characterization factor (Chapter 7):

$$\text{Watson K} = (T_{\text{B}})^{1/3}/\text{sp. gr.}$$

where T_{B} is the absolute boiling point in degrees Rankine and sp. gr. is specific gravity at 15.6°C (60°F). Petroleum, having a wide boiling range, may be described as having several characterization factors for specific boiling fractions of the oil, and the factors for the different fractions may be combined in the correct proportions to give an overall factor for the crude oil.

A highly paraffinic crude oil might have a characterization factor as high as 13 whereas the factor for a highly naphthenic crude oil could be as low as about 10.5. Highly paraffinic crude oils can also contain heavy waxes, which make it difficult for the oil to flow. Thus, another test for paraffin content is to determine how cold a crude oil can be before it fails to flow under specific test conditions. The higher the pour point temperature, the greater the paraffin content for a given boiling range. Once a refinery feedstock has been defined sufficiently well to indicate the primary methods of refining, the nature of the processing operations dictates the product slate.

The characterization factor might also be used as a guide to estimate the potential for instability or incompatibility when feedstocks are blended before entry into the refinery. And instability or incompatibility during refining operations is real and must be addressed during processing.

Instability or *incompatibility*, resulting in the separation of solids during refining, can occur during a variety of process, either by intent (such as in the deasphalting process) or inadvertently when the separation is detrimental to the process. Thus, separation of solids occurs whenever the solvent characteristics of the liquid phase are no longer adequate to maintain polar and/or high molecular weight material in solution.

Examples of such occurrences are:

1. Asphaltene separation, which occurs when the paraffin content or character of the liquid medium increases;
2. Wax separation, which occurs when there is a drop in temperature or the aromatic content or character of the liquid medium increases;
3. Sludge or sediment formation in a reactor, which occurs when the solvent characteristics of the liquid medium change so that asphaltic or wax materials separate;
4. Coke formation, which occurs at high temperatures and commences when the solvent power of the liquid phase is not sufficient to maintain the coke precursors in solution; and
5. Sludge or sediment formation in fuel products, which occurs because of the interplay of several chemical and physical factors.

Thus, there is a need to define adequately the nature of a refinery feedstock before use as well as the nature of the products in order to anticipate the onset of instability or incompatibility. The occurrence of such phenomena during thermal processes will be discussed in this chapter as it relates to the thermal chemistry of crude oil constituents. The instability or incompatibility of petroleum products is dealt with elsewhere in this book (Chapter 8).

2.0 PROCESSES

When petroleum is heated to temperatures in excess of 350°C (660°F), the rate of thermal decomposition of the constituents increases significantly. The higher the temperature, the shorter the time to achieve a given conversion, and the *severity* of the process conditions is a combination of the residence time of the crude oil constituents in the reactor and the temperature needed to achieve a given conversion.

Thermal conversion does not require the addition of a catalyst. This approach is the oldest technology available for residue conversion, and the severity of thermal processing determines the conversion and the product characteristics. As the temperature and residence time are increased, the primary products undergo further reaction to produce various secondary products, and so on, with the ultimate products (coke and methane) being formed at extreme temperatures of approximately 1000°C (1830°F).

Thermal cracking processes are commonly used to convert petroleum vacuum residua into distillable liquid products, although thermal cracking processes as used in the early refineries are no longer in use. Examples of modern thermal cracking processes are *visbreaking* and *coking* (*delayed coking, fluid coking*, and *flexicoking*) (Chapter 6). Although the objective of the visbreaking process is to produce usable liquid products and not to produce coke, sediment is often formed, and in all of these processes the simultaneous formation of sediment or coke limits the conversion to usable liquid products. More recently, the concept of *ultrapyrolysis* (which employs temperatures in excess of 1000°C, 1830°F) for complete conversion to olefins and light ends has been applied to crude oil conversion, but it has not yet seen wide acceptance as a refinery process (Speight, 1991; Gray, 1994).

It is often assumed that, if no side reactions occur, longer residence times at a lower temperature are equivalent to shorter residence times at a higher temperature. However, this assumption does not acknowledge the change in thermal chemistry that can occur at the higher temperatures, irrespective of the residence time. Thermal conditions can, indeed, induce a variety of different reactions in crude oil constituents, so that selectivity for a given product may change considerably with temperature. The onset of secondary, tertiary, and even quaternary reactions under the more extreme high-temperature conditions can convert higher molecular weight constituents of petroleum to low-boiling distillates, butane, propane, ethane, and (ultimately) methane. Caution is advised in the use of extreme temperatures.

However, before progressing to the thermal processes, there is one process in which incompatibility is used to a beneficial effect in product separation, and that is the deasphalting process.

2.1 Visbreaking

To study the thermal chemistry of petroleum constituents, it is appropriate to select the visbreaking process (a carbon rejection process) and the hydrocracking process (a hydrogen addition process) as used in an integrated refinery (Figure 5-1). The processes operate under different conditions (Figure 5-3) and have different levels of conversion (Figure 5-4) and, although they do offer different avenues for conversion, these processes are illustrative of the thermal chemistry that occurs in refineries.

The visbreaking process is primarily a means of reducing the viscosity of heavy feedstocks by *controlled thermal decomposition* insofar as the hot products are quenched before complete conversion can occur (Speight, 1991). However, the process is often plagued by sediment formation in the products. This sediment, or sludge, must be removed if the products are to meet fuel oil specifications.

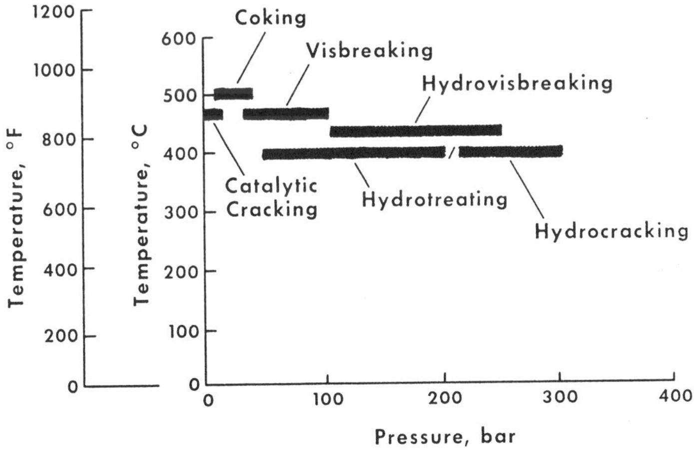

Figure 5-3. Temperature and pressure ranges for conversion and upgrading processes.

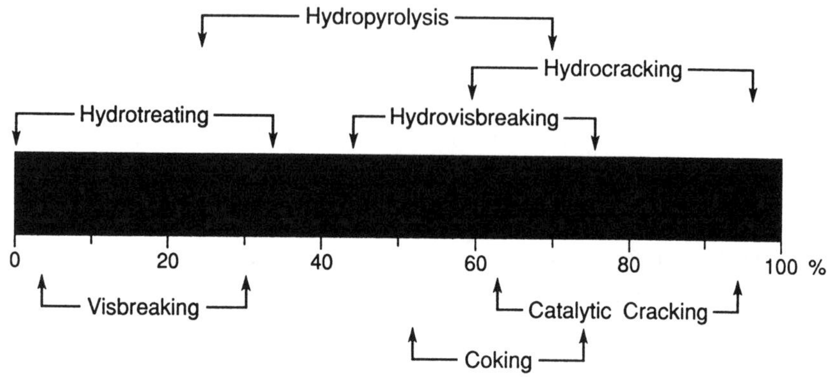

Figure 5-4. Extent of feedstock conversion in various processes.

The process (Figure 5-5) uses the mild thermal cracking (*partial conversion*) as a relatively low-cost and low-severity approach to improving the viscosity characteristics of the residue without attempting significant conversion to distillates. Low residence times are required to avoid polymerization and coking reactions, although additives can help to suppress coke deposits on the tubes of the furnace (Allan et al., 1983).

A visbreaking unit consists of a reaction furnace, followed by quenching with a recycled oil, and fractionation of the product mixture. All of the reaction in this process occurs as the oil flows through the tubes of the reaction furnace. The severity is controlled by the flow rate through the furnace and the temperature; typical conditions are 475–500°C (885–930°F) at the furnace exit with a residence time of 1–3 minutes, and operation for 3–6 months on stream (continuous use) is possible before the furnace tubes must be cleaned and the coke removed (Gary and Handwerk, 1984). The operating pressure in the furnace tubes can range from 0.7–5 MPa depending on the degree of vaporization and the residence time desired. For a given furnace tube volume, a lower operating pressure will reduce the actual residence time of the liquid phase.

The reduction in viscosity of the unconverted residue tends to reach a limiting value with conversion, although the total product viscosity can continue to decrease (Gray, 1994). Conversion of residue in visbreaking follows first-order

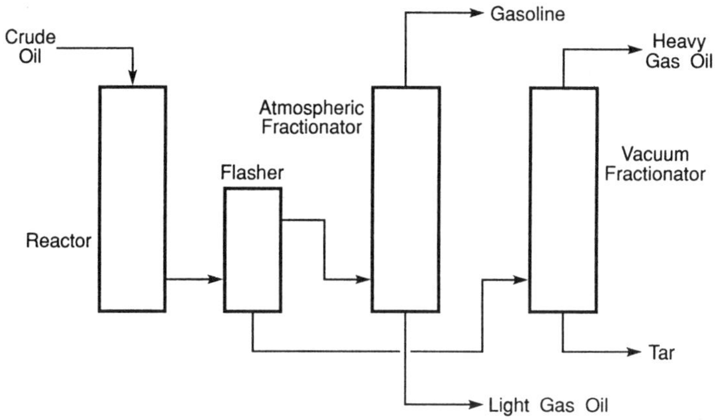

Figure 5-5. The visbreaking process.

reaction kinetics (Henderson and Weber, 1965). The minimum viscosity of the unconverted residue can lie outside the range of allowable conversion if sediment begins to form (Rhoe and de Blignieres, 1979). When pipelining of the visbreaker product is a process objective, a diluent such as gas condensate can be added to achieve a further reduction in viscosity.

The high viscosity of the heavier feedstocks and residua is thought to be due to entanglement of the high molecular weight components of the oil and the formation of ordered structures in the liquid phase. Thermal cracking at low conversion can remove side chains from the asphaltenes and break bridging aliphatic linkages. A 5–10% conversion of atmospheric residue to naphtha is sufficient to reduce the entanglements and structures in the liquid phase and give at l east a five-fold reduction in viscosity. Reduction in viscosity is accompanied by a reduction in the pour point.

The stability of visbroken products is also an issue that might be addressed at this time. Using this simplified model, visbroken products might contain polar species that have been denuded of some of the alkyl chains and which, on the basis of solubility, might be more rightly called *carbenes* and *carboids*, but an induction period is required for phase separation or agglomeration to occur. Such products might initially be *soluble* in the liquid phase but after the induction period, cooling, and/or diffusion of the products, incompatibility (phase separation, sludge formation, agglomeration) occurs.

2.2 Coking

Coking, as the term is used in the petroleum industry, is a process for converting nondistillable fractions (residua) of crude oil to lower boiling products and coke. Coking is often used in preference to catalytic cracking because of the presence of metals and nitrogen components that poison catalysts.

There are several coking processes: *delayed coking*, *fluid coking*, and *flexicoking*, as well as several other variations.

Delayed coking (Figure 5-6) is the oldest, most widely used process and has changed very little in the five or more decades in which it has been on stream in refineries (Speight, 1991).

In the semicontinuous process, the residuum or other heavy feedstock is heated to the cracking/coking temperature (>350°C, >660°F; but usually at temperatures on the order of 480°C, 895°F), and the hot liquid is charged, usually by upflow, to the coke drum, where the coking reactions occur.

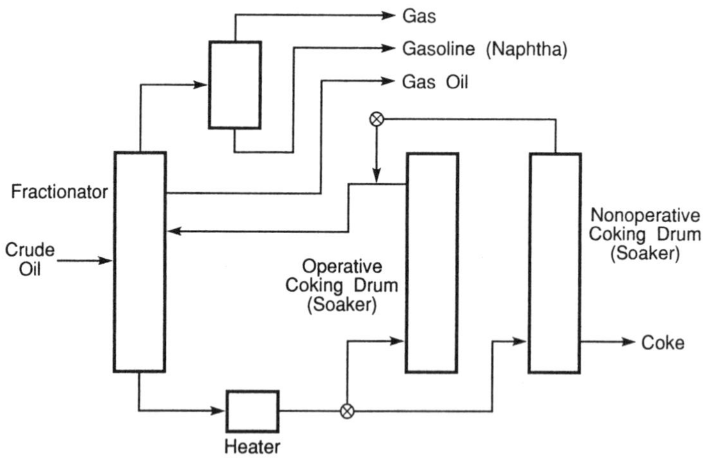

Figure 5-6. The delayed coking process.

Liquid and gaseous products pass to the fractionator for separation and coke deposits in the drum. The coke drums are arranged in pairs, one on stream and the other off stream, and used alternately to allow continuous processing. The process can be operated on a cycle, typically 24–48 hours.

The overhead oil is fractionated into fuel gas (ethane and lower molecular weight gases), propane-propylene, butane-butene, naphtha, light gas oil, and heavy gas oil. Yields and product quality vary widely due to the broad range of feedstock types charged to delayed coking. The function of the coke drum is to provide the residence time required for the coking reactions and to accumulate the coke. Hydraulic cutters are used to remove coke from the drum.

Fluid coking (Figure 5-7) is a continuous fluidized solids process that cracks feed thermally over heated coke particles in a reactor vessel to gas, liquid products, and coke. Heat for the process is supplied by partial combustion of the coke, with the remaining coke being drawn as product. The new coke is deposited in a thin fresh layer on the outside surface of the circulating coke particle.

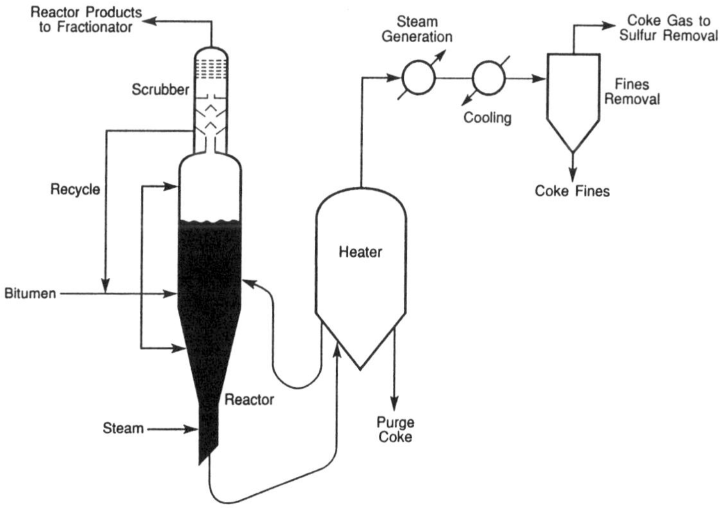

Figure 5-7. The fluid coking process.

Small particles of coke made in the process circulate in a fluidized state between the vessels and are the heat transfer medium. Thus the process requires no high-temperature preheat furnace. Fluid coking is carried out at essentially atmospheric pressure and temperatures in excess of 485°C (900°F) with residence times of the order of 15–30 seconds. The longer residence time is in direct contrast to the delayed coking process, in which the coking reactions are allowed to proceed to completion. This is evident from the somewhat higher liquid yields observed in many fluid coking processes. However, the products from a fluid coker may be somewhat more olefinic and slightly less desirable for downstream processing.

The *flexicoking* process (Figure 5-8) is a modification of the fluid coking process that includes a gasifier adjoining the burner/regenerator to convert excess coke to a clean fuel gas with a heating value of about 90 Btu/ft^3. The coke gasification can be controlled to burn about 95% of the coke to maximize production of coke gas or to operate at a reduced level to produce both gas and a coke. This flexibility permits adjustment for coke market conditions over a considerable range of feedstock properties.

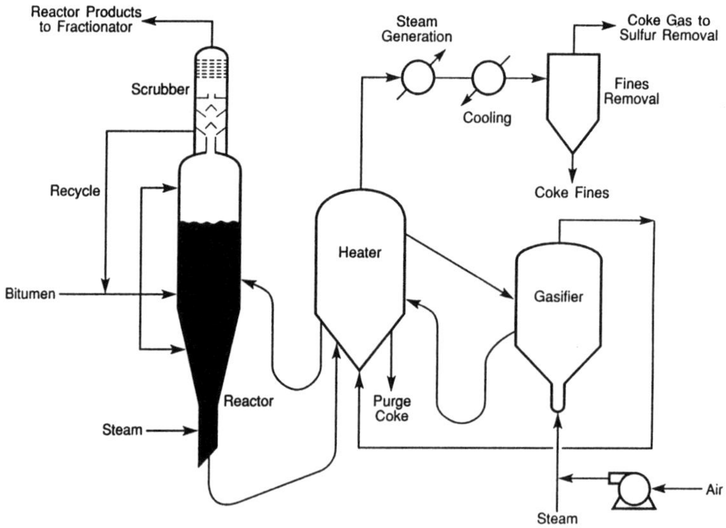

Figure 5-8. The flexicoking process.

The *liquid products* from the coker can, following cleanup via commercially available hydrodesulfurization technology (Speight, 1981), provide low-sulfur liquid fuels (less than 0.2% w/w sulfur). Coker naphthas have boiling ranges up to 220°C (430°F), are olefinic, and must be upgraded by hydrogen processing for removal of olefins, sulfur, and nitrogen. They are then used conventionally for reforming to gasoline or chemicals feedstock. Middle distillates, boiling in the range 220–360°C (430–680°F), are also hydrogen treated for improved storage stability, sulfur removal, and nitrogen reduction. They can then be used as precursors to gasoline, diesel fuel, or fuel oil. The gas oil boiling up to 510°C (950°F) is usually low in metals and may be used as the feedstock for fluid catalytic cracking.

Another major application for the coking processes is in upgrading heavy (high-viscosity) low-value crude oils into lighter products.

Petroleum *coke* is used principally as a fuel or, after calcining, for carbon electrodes. The feedstock from which the coke is produced controls the coke properties, especially the sulfur, nitrogen and metals content. A concentration effect tends to deposit the majority of the sulfur, nitrogen, and metals in the coke. Cokes exceeding about 2.5% sulfur content and 200 ppm vanadium are mainly used, environmental regulations permitting, for fuel or fuel additives. The properties of coke for nonfuel use include a low sulfur, metals, and ash content as well as a definable physical structure.

2.3 Hydrotreating and Hydrocracking

Hydrotreating is defined as the lower temperature removal of heteroatomic species by treatment of a feedstock or product in the presence of hydrogen. Hydrocracking (Figure 5-9) is the thermal decomposition of a feedstock in which carbon-carbon bonds are cleaved in addition to the removal of heteroatomic species. Hydrogen is present to prevent the formation of coke.

Thus, the presence of hydrogen changes the nature of the products, especially the coke yield, by preventing the buildup of precursors that are incompatible in the liquid medium and form coke (Magaril and Aksenova, 1968; Magaril and Ramazaeva, 1969; Magaril and Aksenova, 1970; Magaril et al., 1970; Speight and Moschopedis, 1979). In fact, the chemistry involved in the reduction of asphaltenes to liquids using models in which the polynuclear aromatic system borders on graphitic is difficult to visualize. However, the *paper chemistry* derived from the use of a molecularly designed model composed of smaller polynuclear aromatic systems is much easier to visualize (Speight, 1991). But precisely how asphaltenes react with the catalysts is open to much more speculation.

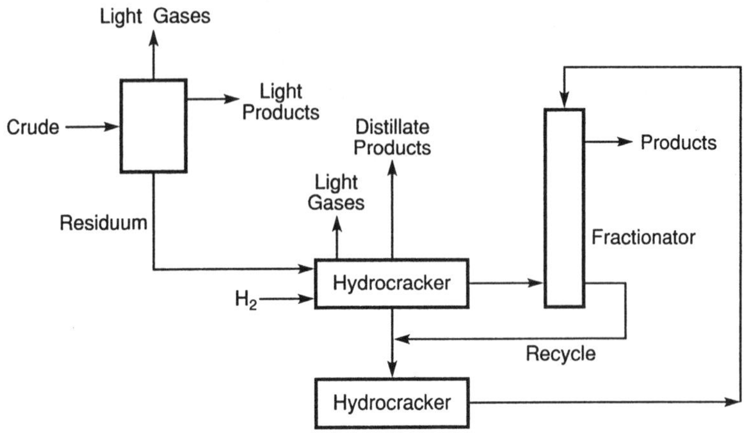

Figure 5-9. The hydrocracking process.

Deposition of solids or incompatibility is still possible when asphaltenes interact with catalysts, especially acidic support catalysts, through the functional groups, e.g., the basic nitrogen species just as they interact with adsorbents. And there is a possibility for interaction of the asphaltene with the catalyst through the agency of a single functional group in which the remainder of the asphaltene molecule remains in the liquid phase. There is also a less desirable option in which the asphaltene reacts with the catalyst at several points of contact, causing immediate incompatibility on the catalyst surface.

3.0 PROCESS CHEMISTRY AND PHYSICS

Thermal processes are the oldest methods for crude oil conversion and are still used in modern refineries. The thermal chemistry of petroleum constituents has been investigated for more than five decades, and the precise chemistry of the lower molecular weight constituents has been well defined because of the bountiful supply of pure compounds. The major issue in determining the thermal chemistry of the nonvolatile constituents is, of course, their largely unknown chemical nature and, therefore, inability to define their thermal chemistry with any degree of accuracy. Indeed, it is only recently that some light has been cast on the thermal chemistry of the nonvolatile constituents.

The nonvolatile constituents of petroleum generally produce coke in yields varying from almost zero to more than 60 wt. % (Figure 5-10). However, the focus of thermal studies has been, for obvious reasons, on the apshaltenes, which produce thermal coke in amounts varying from approximately 35 to approximately 65 wt. %. Petroleum mapping techniques often show the nonvolatile constituents, specifically the asphaltenes and the resins, producing coke while the volatile constituents produce distillates. It is often ignored that the asphaltenes also produce high yields (35–65 wt. %) of volatile thermal products, which vary from condensible liquids to gases.

Thus, the challenges facing process chemistry and physics are determining (1) the means by which petroleum constituents thermally decompose, (2) the nature of the products of thermal decomposition, (3) the subsequent decompositon of the *primary* thermal products, (4) the interaction of the products with each other, (5) the interaction of the products with the original costituents, and (6) the influence of the products on the composition of the liquids.

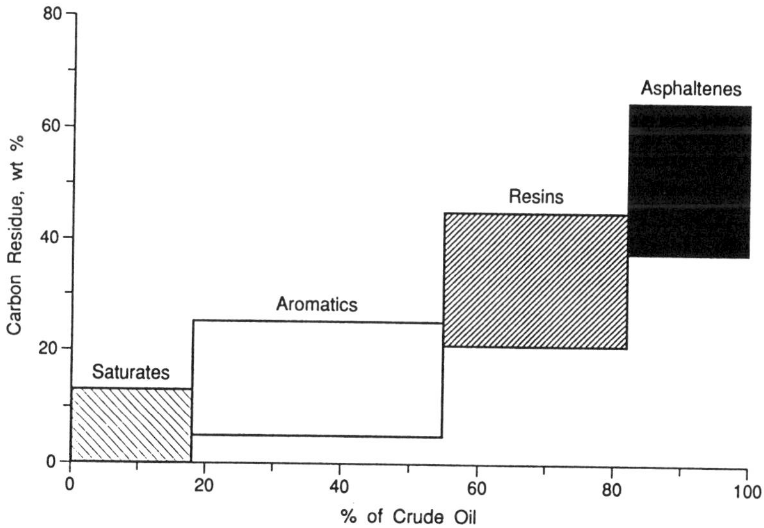

Figure 5-10. Coke formation from various fractions.

When petroleum is heated to temperatures over approximately 410°C (770°F), the thermal or free radical reactions start to crack the mixture at significant rates. Thermal conversion does not require the addition of a catalyst; therefore, this approach is the oldest technology available for residue conversion. The severity of thermal processing determines the conversion and the product characteristics.

As asphaltenes are a major component of residua and heavy oils, their thermal decomposition has been the focus of much attention. The thermal decomposition not only produces high yields (40 wt. %) of coke but also, optimistically and realistically, produces equally high yields of volatile products (Speight, 1970). Thus, the challenge in studying the thermal decomposition of asphaltenes is to decrease the yields of coke and increase the yields of volatile products.

Several chemical models (Wiehe, 1993a, 1993b and references cited therein; Gray, 1994 and references cited therein; Speight, 1994 and references cited therein) describe the thermal decomposition of asphaltenes. These can be described in composite form and are worthy of consideration. Using these available asphaltene models as a guide, the prevalent thinking is that the asphaltene nuclear fragments become progressively more polar as the paraffinic fragments are stripped from the ring systems by scission of the bonds (preferentially) between the carbon atoms alpha and beta to the aromatic rings.

The polynuclear aromatic systems that have been denuded of the attendant hydrocarbon moieties are somewhat less soluble in the surrounding hydrocarbon medium than their *parent* systems (Bjorseth, 1983; Dias, 1987, 1988) not only because of the solubilizing alkyl moieties but also because of the enrichment of the liquid medium in paraffinic constituents. Again, there is an analogy with the deasphalting process (Chapter 8), except that the paraffinic material is a product of the thermal decomposition of the asphaltene molecules and is formed in situ rather than being added separately.

In contrast to the visbreaking process, in which the general principle is the production of products for use as fuel oil, the hydrocracking process is employed to produce a slate of products for use as liquid fuels. Nevertheless, the decomposition of asphaltenes is, again, an issue, and just as models consisting of large polynuclear aromatic systems are inadequate to explain the chemistry of visbreaking, they are also of little value for explaining the chemistry of hydrocracking.

However, it must be recognized that coke is not the only product, and a simple weight balance shows that volatile materials can be the major products of the thermal treatment of the asphaltenic fractions of crude oil (Figure 5-11).

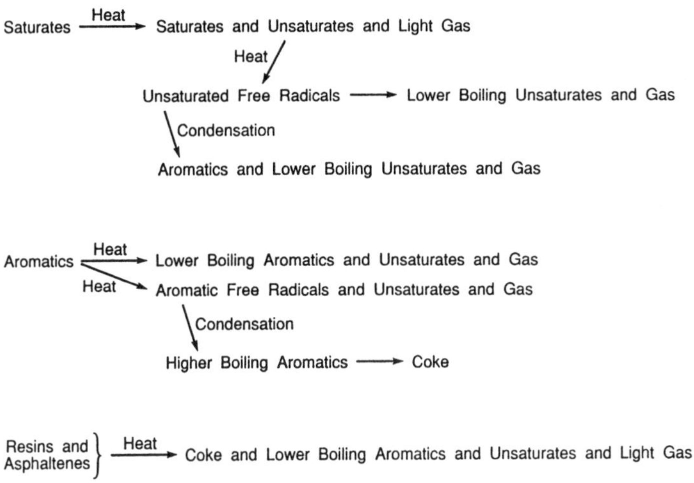

Figure 5-11. Representation of the products from the various fractions of petroleum.

The coke has a lower hydrogen-to-carbon atomic ratio than any of the constituents originally present in the crude oil, and the hydrocarbon products *may* have higher hydrogen-to-carbon atomic ratios than any of the constituents originally present in the crude oil or hydrogen-to-carbon atomic ratios at least equal to those of many of the original constituents. It must also be recognized that the production of coke and volatile hydrocarbon products is accompanied by a shift in the hydrogen distribution (Speight, 1991).

The initial stages of the thermal decomposition of asphaltenes have been least well understood. It is known that the overall thermal decomposition involves scission of carbon-carbon bonds, aromatization of naphthenic ring systems, formation of lower molecular weight material, and formation of coke. The presence of hydrogen during the thermal decomposition of asphaltenes can alter the product slate. However, the chemistry remains complex, and representation of the chemistry of asphaltene thermolysis is speculative and has been done simply by the use of nonchemical formulae (Speight, 1991).

Thermal treatment of residua ranges from mild treatment for reduction of viscosity to extremely high temperature treatment (>1000°C, >1830°F) for complete conversion to olefins and light ends. The higher the temperature, the shorter the time to achieve a given conversion. The severity of the process conditions is the combination of reaction time and temperature to achieve a given conversion.

If no side reactions occur, very long residence times at low temperature should be equivalent to very short times at high temperature. Thermal reactions, however, can give rise to a variety of different reactions, so selectivity for a given product changes with temperature and pressure. In fact, the nature of the thermal processes varies considerably.

Mild- and high-severity processes are frequently used for processing of residue fractions, whereas conditions similar to those of *ultrapyrolysis* (high temperature and very short residence time) are used commercially only for cracking ethane, propane, butane, and light distillate feeds to produce ethylene and higher olefins.

The formation of solid sediments, or coke, during thermal processes is a major limitation on processing. Furthermore, the presence of different types of solids shows that solubility controls the formation of solids. And the tendency for solid formation changes in response to the relative amounts of the light ends, middle distillates, and residues and to their changing chemical composition during the process (Gray, 1994). In fact, the prime mover in the formation of incompatible products during the processing of feedstocks containing asphaltenes is the nature of the primary thermal decomposition products, particularly those designated as *carbenes* and *carboids* (Chapter 1) (Speight, 1987; Speight, 1992; Wiehe, 1992; Wiehe, 1993a, 1993b; Calemma et al., 1994).

Coke formation during the thermal treatment of petroleum residua is postulated to occur by a mechanism that involves the liquid-liquid phase separation of reacted asphaltenes (which may be *carbenes*) to form a phase that is lean in abstractable hydrogen. The unreacted asphaltenes were found to be the fraction with the highest rate of thermal reaction but with the least extent of reaction. This not only described the appearance and disappearance of asphaltenes but also quantitatively described the variation in molecular weight and hydrogen content of the asphaltenes with reaction time. Further evidence for the liquid-liquid phase separation was the observation of spherical particles of liquid crystals.

One of the postulates of coke formation involves the production of coke by a sequence of polymerization and condensation steps from the lightest (lower molecular weight) to the heaviest (higher molecular weight) fractions. Present evidence (Speight, 1991, 1994, Wiehe, 1992, 1993a, 1993b) shows that the reaction pathway of the asphaltenes to coke is much more complex and involves at least several reaction stages (Figure 5-12).

In addition, there is an induction period during which no coke is formed, and many kinetic models that contain sequences of direct chemical reactions to coke fail to predict the induction period. This induction period has been observed experimentally by many previous investigators (Levinter et al., 1966, 1967; Magaril and Aksenova, 1968; Magaril and Aksenova, 1970; Valyavin et al., 1979; Takatsuka et al., 1989a) and makes visbreaking and the Eureka processes possible. The postulation that that coke formation is triggered by the phase separation of asphaltenes (Magaril et al., 1971) led to the use of linear variations of the concentration of each fraction with reaction time, resulting in the assumption of zero-order kinetics rather than first-order kinetics. More recently (Yan, 1987), coke formation in visbreaking was described as resulting from a phase separation step, but the phase-separation step was not included in the resulting kinetic model for coke formation.

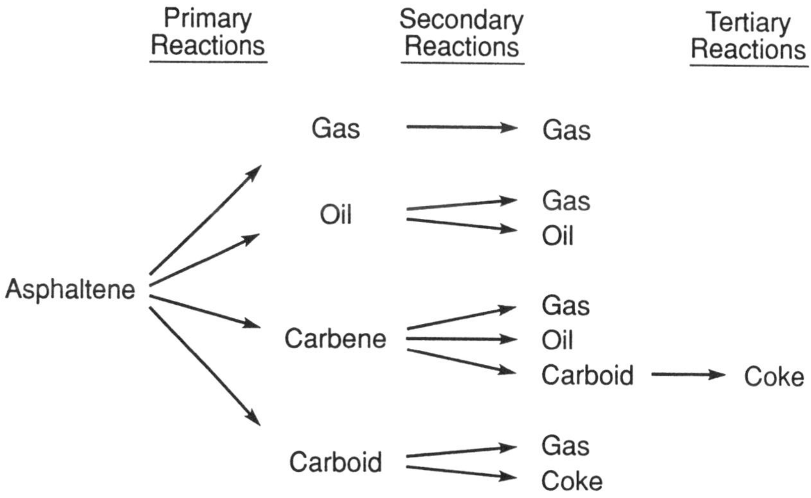

Figure 5-12. Representation of the multistep thermal reactions of asphaltenes.

This mechanism also appears in studies of residua hydroconversion, which has included a phase-separation step (the formation of *dry sludge*) in a kinetic model but was not included as a preliminary step to coke formation in a thermal cracking model (Sosnowski et al., 1980; Takatsuka et al., 1989a, 1989b).

Several kinetic studies have concentrated on the conversion of the asphaltene portion of petroleum residua. One in particular (Schucker and Keweshan, 1980) showed that asphaltenes exhibit a coke induction period when thermally converted at 400° C (750°F), but the kinetic model did not include a coke formation step. Other workers (Savage et al., 1985, 1988) also observed a coke-induction period for the thermolysis of asphaltenes that disappeared when the thermolysis temperature was raised to 450°C (840°F). It was proposed (Savage and Klein, 1989) that toluene-insoluble coke could be defined as a molecular species with a molecular weight greater than 300 and an hydrogen-to-carbon atomic ratio less than 1.

It has also been shown (Wiehe, 1992) that the use of molecular weight and the hydrogen-to-carbon atomic ratio to assign solubility fractions is reasonable. A regular-solution model has been used for liquid-solid equilibrium prediction to determine how species are divided into maltenes, asphaltenes, and coke (Neurock et al., 1990; Nigam et al., 1990).

More recently (Wiehe, 1993a, 1993b), a kinetic model was developed that describes four common features of residua thermal conversion kinetics: (1) an induction period prior to coke formation, (2) a maximum concentration of asphaltenes in the reacting liquid, (3) a decrease in the asphaltene concentration that parallels the decrease in heptane-soluble material, and (4) the high reactivity of the unconverted asphaltenes.

This model represents the conversion of asphaltenes over the entire temperature range and of heptane-soluble materials in the coke induction period as first-order reactions. The data also show that the four reactions give simultaneously lower aromatic and higher aromatic products, on the basis of other evidence (Wiehe, 1992). Also, the previous work showed that residua fractions can be converted without completely changing solubility classes. This is represented by having both reactant and product species for asphaltenes and heptane-soluble materials and uses the concept (Magaril et al., 1971) that coke formation is triggered by the phase separation of converted asphaltenes.

The maximum solubility of these product asphaltenes is proportional to the total heptane-soluble materials, as suggested by the observation that the decrease in asphaltenes parallels the decrease of heptane-soluble materials. Finally, the conversion of the insoluble product asphaltenes into toluene-insoluble coke is pictured as producing a heptane-soluble by-product, which provides a mechanism for the heptane-soluble conversion to deviate from first-order behavior once coke begins to form. An infinite reaction rate for this coke-forming reaction is used to show that the reaction rate is phase equilibrium controlled:

$$H^+ = aA^* + (1-a)V \qquad \text{reaction constant } k_H$$

$$A+ = mA^* + nH^* + (1-m-n)V \qquad \text{reaction constant } k_A$$

$$\text{solubility limit: } A^*_{max} = S_L(H^+ + H^*)$$

$$A^*_{ex} = A^* - A^*_{max}$$

$$A^*_{ex} = (1-y)TI + yH^*$$

where a is a stoichiometric coefficient; A^+, reactant asphaltenes; A*, asphaltene cores; A^*_{max}, maximum asphaltene cores that can be held in solution; A^*_{ex}, excess asphaltene cores beyond what can be held in solution; H^+, reactant, nonvolatile heptane-soluble materials; H*, product, nonvolatile heptane-soluble materials; k_A first-order reaction rate constant for reactant asphaltene thermolysis (min^{-1}); k_H, first-order reaction rate constant for the thermolysis of reactant heptane-soluble materials (min^{-1}); m, stoichiometric coefficient; n, stoichiometric coefficient; S_L, solubility limit (wt %/wt %); TI, toluene-insoluble coke; and V, volatile materials.

The first two parallel first-order reactions for the thermolysis of unreacted heptane-soluble materials and unreacted asphaltenes are the only reactions that occur during the coke induction period. During the coke induction period the reactant asphaltenes form only lower molecular weight products and, as long as the asphaltenes remain dissolved, the heptane-soluble materials can provide sufficient abstractable hydrogen to terminate asphaltene free radicals, making asphaltene radical–asphaltene radical recombination infrequent.

In support of this assumption, it is known (Langer et al., 1961) that partially hydrogenated refinery process streams provide abstractable hydrogen and, as a result, inhibit coke formation during residuum thermal conversion. Thus, the heptane-soluble fraction of a residuum, which contains naturally occurring, partially hydrogenated aromatics, can provide abstractable hydrogen during thermal reactions.

As the conversion proceeds, the concentration of asphaltene cores continues to increase and the heptane-soluble fraction continues to decrease until the solubility limit, S_L, is reached. Beyond the solubility limit, the excess asphaltene cores, A^*_{ex}, phase separate to form a second liquid phase that is lean in abstractable hydrogen. In this new phase asphaltene radical–asphaltene radical recombination is quite frequent, causing a rapid reaction to form solid coke and a by-product of a heptane-soluble core.

The asphaltene concentration varies little in the coke induction period (Wiehe, 1993a, 1993b) but decreases once coke begins to form. Observing this, it might be concluded that asphaltenes are unreactive, but it is the high reactivity of the asphaltenes down to the asphaltene core that offsets the generation of asphaltene cores from the heptane-soluble materials to keep the overall asphaltene concentration nearly constant.

Previously, it was demonstrated (Schucker and Keweshan, 1980; Savage et al., 1988) that the hydrogen-to-carbon atomic ratio of the asphaltenes decreases rapidly with reaction time for asphaltene thermolysis and then approaches an asymptotic limit at long reaction times, which provides qualitative evidence for asphaltene cracking down to a core.

The measurement of the molecular weight of petroleum asphaltenes is known to give different values depending on the technique, the solvent and the temperature (Chapter 4) (Dickie and Yen, 1967; Moschopedis et al., 1976; Speight et. al., 1985). As shown by small-angle x-ray (Kim and Long, 1979) and neutron (Overfield et al., 1989) scattering, this is because asphaltenes tend to self-associate and form aggregates.

There is a need to investigate fully the degree of association of asphaltenes in thermal systems and to determine whether it has any influence on the coke-forming tendencies of the asphaltenes. The *association factor*, I (Wiehe, 1993a, 1993b), is the ratio of the associated molecular size of the asphaltene measured in toluene, the poorest solvent required to dissolve the asphaltene, at a relatively low temperature divided by a measure of an *unassociated* molecular weight (Wiehe, 1992).

The data for the thermolysis reactions (Wiehe, 1993a, 1993b) indicated that the association factor is nearly constant at approximately two, but when coke formation commences the remaining asphaltenes have less tendency to associate. As more and more toluene-insoluble materials are formed, the association factor continually decreases until the remaining asphaltenes show almost no tendency to associate (I = 1). However, there is evidence that even if association occurs at the high temperatures involved in thermal decomposition, it may not make any difference to the chemistry of coke formation (Rao and Serrano, 1986; Schabron and Speight, 1997).

Thus, the available data suggest that coke formation is a complex process involving both chemical reactions and thermodynamic behavior. Reactions that contribute to this process are cracking of side chains from aromatic groups. dehydrogenation of naphthenes to form aromatics, condensation of aliphatic structures to form aromatics, condensation of aromatics to form higher fused-ring aromatics, and dimerization or oligomerization reactions. Loss of side chains always accompanies thermal cracking, and dehydrogenation and condensation reactions are favored by hydrogen-deficient conditions.

The importance of solvents in coking has been recognized for many years (e.g. Langer et al., 1961), but their effects have often been ascribed to hydrogen donor reactions rather than phase behavior. The separation of the phases depends on the solvent characteristics of the liquid. Addition of aromatic solvents suppresses phase separation, whereas paraffins enhance separation. Microscopic examination of coke particles often shows evidence for mesophases, spherical domains that exhibit the anisotropic optical characteristics of liquid crystals.

This phenomenon is consistent with the formation of a second liquid phase; the mesophase liquid is denser than the rest of the hydrocarbon, has a higher surface tension, and probably wets metal surfaces better than the rest of the liquid phase. The mesophase characteristic of coke diminishes as the liquid phase becomes more compatible with the aromatic material (Lott and Cyr, 1992).

The phase separation phenomenon that is the prelude to coke formation can also be explained by use of the solubility parameter, Λ, for petroleum fractions and for the solvents. Although little is known about the solubility parameter of petroleum fractions, there has been a noteworthy attempt to define the solubility parameter ranges for different fossil fuel liquids (Yen, 1984; Speight, 1994).

As an extension of this concept, there is sufficient fragmented data to draw an approximate correlation between H/C atomic ratio and the solubility parameter for hydrocarbons and the constituents of the lower boiling fractions of petroleum (Speight, 1994). Hydrocarbon liquids can dissolve polynuclear hydrocarbons, a case in which there is usually less than a three-point difference between the lower solubility parameter of the solvent and the higher solubility parameter of the solute. Thus, a parallel, or near-parallel, line can be assumed that allows the solubility parameter of the asphaltenes and resins to be estimated.

By this means, the solubility parameter of asphaltenes can be estimated to fall in the range 9–12, which is in keeping with the asphaltenes being composed of a mixture of different compound types with an accompanying variation in polarity. Removal of alkyl side chains from the asphaltenes decreases the hydrogen-to-carbon atomic ratio (Speight, 1991; Wiehe, 1993a, 1993b; Gray, 1994) and increases the solubility parameter, thereby bringing about a concurrent decrease of the asphaltene product in the hydrocarbon solvent.

In fact, on the molecular weight polarity diagram for asphaltenes, carbenes and carboids can be shown as lower molecular weight, highly polar entities in keeping with molecular fragmentation models (Speight, 1991, 1994). If this increase in polarity and solubility parameter (Mitchell and Speight, 1973) is too drastic relative to the surrounding medium (Figures 5-13 and 5-14), phase separation will occur.

Any chemical or physical interaction that causes a change in the solubility parameter of the solute relative to that of the solvent will also cause incompatibility be it called instability, incompatibility, phase separation, or sediment/sludge formation.

Several noteworthy attempts have been made to focus attention on the asphaltenes during hydroprocessing studies. The focus has been on the macromolecular changes that occur by investigation of the changes in the generic fractions, i.e., the asphaltenes, the resins, and the other fractions that make up such a feedstock (Drushel, 1972).

This option suggests that the overall pathway by which hydrotreating and hydrocracking of heavy oils and residua occur involves a stepwise mechanism:

Asphaltenes → polar aromatics (resin-type components)

Polar aromatic → aromatics

Aromatics → saturates

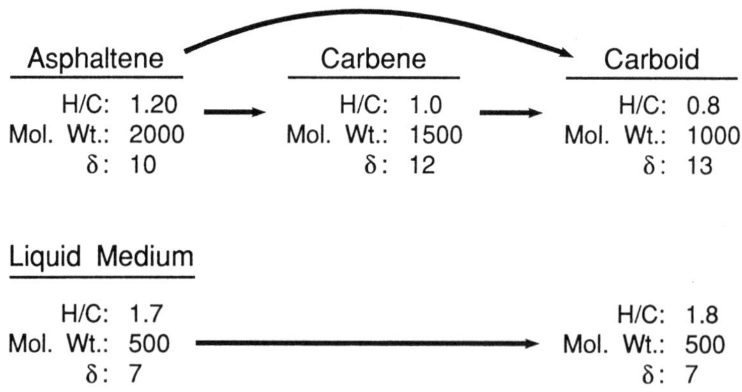

Figure 5-13. Representation of the changes to the maltenes and the asphaltenes during thermal cracking.

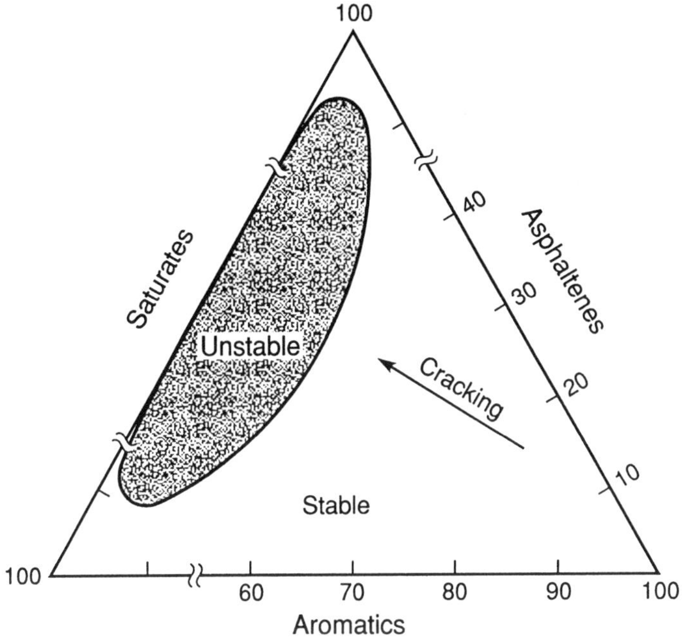

Figure 5-14. Representation of petroleum as a three-phase system to illustrate instability or phase separation as a result of cracking.

A direct step from either the asphaltenes or the resin-type components to the saturates is not considered a predominant pathway for hydroprocessing.

The means by which asphaltenes are desulfurized, as one step of a hydrocracking operation, is also suggested as part of this process. This concept can then be taken one step further to show the dealkylation of the aromatic systems as a definitive step in the hydrocracking process (Speight, 1987). It is also likely that molecular species within the asphaltene fraction that contain nitrogen and other heteroatoms, and have lower volatility than their hydrocarbon analogs, are the prime movers in the production of coke (Speight, 1987).

When catalytic processes are employed, complex molecules, such as those that may be found in the original asphaltene fraction or those formed during the process, are not sufficiently mobile (or are too strongly adsorbed by the catalyst) to be saturated by the hydrogenation components and, hence, continue to condense and eventually degrade to coke. These deposits deactivate the catalyst sites and eventually interfere with the hydroprocess.

A convenient means of understanding the influence of feedstock on the hydrocracking process is through a study of the hydrogen content (hydrogen-to-carbon atomic ratio) and molecular weight (carbon number) of the feedstocks and products (O'Rear et al., 1982). Such data show the extent to which the carbon number must be reduced and/or the relative amount of hydrogen that must be added to generate the desired lower molecular weight, hydrogenated products. In addition, it is possible to use data for hydrogen usage in residuum processing, where the relative amount of hydrogen consumed in the process can be shown to be dependent upon the sulfur content of the feedstock (Bridge et al., 1975).

In summary, and as these concepts show, characterization data can be used as an integral part of understanding the thermal chemistry of asphaltenes as a start in predicting product yield and product distribution; predictions of product quality will follow.

4.0 REFERENCES

Allan, D. E., Martinez, C. H., Eng. C. C., and Barton, W.J. 1983. Chem. Eng. Progr. 79(1): 85.

Bjorseth, A. 1983. Handbook of Polycyclic Aromatic Hydrocarbons. Marcel Dekker Inc., New York.

Dias, J. R. 1987. Handbook of Polycyclic Hydrocarbons. Part A. Benzenoid Hydrocarbons. Elsevier, New York.

Dias, J. R. 1988. Handbook of Polycyclic Hydrocarbons. Part B. Polycyclic Isomers and Analogs of Benzenoid Hydrocarbons. Elsevier, New York.

Dickie, J. P., and Yen, T. F. 1967. Anal. Chem. 39: 1847.

Gary, J. G., and Handwerk, G. E. 1984. Petroleum Refining: Technology and Economics. Marcel Dekker Inc., New York.

Gray, M. R. 1994. Upgrading Petroleum Residues and Heavy Oils. Marcel Dekker Inc., New York.

Henderson, J. H., and Weber, L. 1965. J. Can. Pet. Tech. 4: 206.

Kim, H., and Long, R. B. 1979. Ind. Eng. Chem. Fundam. 18: 60.

Langer, A. W., Stewart, J., Thompson, C. E., White, H. T., and Hill, R. M. 1961. Ind. Eng. Chem. 53: 27.

Levinter, M. E., Medvedeva, M. I., Panchenkov, G. M., Aseev, Y. G., Nedoshivin, Y. N., Finkelshtein, G. B., and Galiakbarov, M. F. 1966. Khim. Tekhnol. Topl. Masel. No. 9: 31.

Levinter, M. E., Medvedeva M. I., Panchenkov, G. M., Agapov, G. I., Galiakbarov, M. F., and Galikeev, R. K. 1967. Khim. Tekhol. Topl. Masel. No. 4: 20.

Magaril, R. Z., and Akensova, E. I. 1968. Int. Chem. Eng. 8: 727.

Magaril, R. Z., and Aksenova, E. I. 1970. Khim. Tekhnol. Topl. Masel. No. 7: 22.

Magaril, R. Z., and Ramazaeva, L. F. 1969. Izv. Vyssh. Ucheb. Zaved. Neft Gaz. 12(1): 61.

Magaril, R. Z., Ramazaeva, L. F., and Aksenova, E. I. 1970. Khim. Tekhnol. Topl. Masel. 15(3): 15.

Magaril, R. Z., Ramazeava, L. F., and Aksenora, E. I. 1971. Int. Chem. Eng. 11: 250.

Mitchell, D. L., and Speight, J. G. 1973. Fuel. 52: 149.

Moschopedis, S. E., Fryer, J. F., and Speight, J. G. 1976. Fuel. 55: 227.

Neurock, M., Libanati, C., Nigam, A., and Klein, M. T. 1990. Chem. Eng. Sci. 45: 2083.

Overfield, R. E., Sheu, E. Y., Sinha, S. K., and Liang, K. S. 1989. Fuel Sci. Technol. Int 7: 611.

Rao, B. M. L., and Serrano, J. E. 1986. Fuel Sci. Technol. Int. 4: 483.

Rhoe, A., and de Blignieres, C. 1979. Hydrocarbon Process. 58(1): 131.

Savage, P. E., and Klein, M. T. 1989. Chem. Eng. Sci. 44: 393.

Savage, P. E., Klein, M. T., and Kukes, S. G. 1985. Ind. Eng. Chem. Process Des. Dev. 24: 1169.

Savage, P. E., Klein, M. T., and Kukes, S. G. 1988. Energy & Fuels. 2: 619.

Schabron, J. F., and Speight, J. G. 1997. Preprints Div. Petrol. Chem. Am. Chem. Soc. 42: 386.

Schucker, R. C., and Keweshan, C. F. 1980. Preprints Div. Fuel Chem. Am. Chem. Soc. 25: 155.

Speight, J. G. 1970. Fuel. 49: 134.

Speight, J. G. 1981. The Desulfurization of Heavy Oils and Residua. Marcel Dekker Inc., New York.

Speight, J. G. 1987. Preprints Div. Petrol. Chem. Am. Chem. Soc. 32(2): 413.

Speight, J. G. 1991. The Chemistry and Technology of Petroleum. 2nd Edition. Marcel Dekker Inc., New York.

Speight, J. G. 1992. Proceedings. 4th International Conference on the Stability and Handling of Liquid Fuels. U.S. Department of Energy (DOE/CONF-911102). p. 169.

Speight, J. G. 1994. In Asphalts and Asphaltenes, 1. T.F. Yen and G.V. Chilingarian (Editors). Elsevier, Amsterdam, The Netherlands. Chapter 2.

Speight, J. G., and Moschopedis, S. E. 1979. Fuel Process. Technol. 2: 295.

Speight, J. G., Wernick, D. L., Gould, K. A., Overfield, R. E., Rao, B. M. L., and Savage, D. W. 1985. Rev. Inst. Fr. Petrole. 40: 51.

Takatsuka, T., Kajiyama, R., Hashimoto, H., Matsuo, I., and Miwa, S. A. 1989a. J. Chem. Eng. Jpn. 22: 304.

Takatuska, T., Wada, Y., Hirohama, S., and Fukui, Y. A. 1989b. J. Chem. Eng. Jpn. 22: 298.

Valyavin, G. G., Fryazinov, V. V., Gimaev, R. H., Syunyaev, Z. I., Vyatkin, Y. L., and Mulyukov, S. F. 1979. Khim. Tekhol. Topl. Masel. No. 8: 8.

Wiehe, I. A. 1992. Ind. Eng. Chem. Res. 31: 530.

Wiehe, I. A. 1993a. Preprints Div. Petrol. Chem. Am. Chem. Soc. 38: 428.

Wiehe, I. A. 1993b. Ind. Eng. Chem. Res. 32: 2447.

Yan, T. Y. 1987. Preprints Div. Petrol. Chem. Am. Chem. Soc. 32: 490.

Yen, T. F. 1984. In The Future of Heavy Crude Oil and Tar Sands. R.F. Meyer, J. C. Wynn, and J. C. Olson (Editors). McGraw-Hill, New York.

Chapter 6

Heavy Oil Upgrading Processes

M. Rashid Khan
Texaco Fuels & Lubricants Technology Department
P.O. Box 509
Beacon
New York 12508

and

David J. Patmore
National Center for Upgrading Technology
CANMET Western Research Laboratory
1 Oil Patch Drive
Suite A202
Devon, Alberta
Canada T9G 1A8

1.0 INTRODUCTION

This chapter presents an overview of the current state of the art of upgrading technology. A detailed description of every process is beyond the scope of this chapter, and for this the reader is referred to the several upgrading handbooks available and the references given for each process (Meyers, 1986; Hydrocarbon Process., 1996). Instead, this chapter will discuss upgrading in terms of the type of technology and the chemistry involved.

The need to convert heavy oil and residue is growing. Traditionally, the factor that has affected the growth is economics. The relatively narrow difference in price between light and heavy crude oil historically has deterred the growth of conversion technology. It is often more economic to buy lighter crude oils than to invest in conversion technology.

The current motivations for conversion technologies are as follows:

1. The market for heavy oil and residue is limited. There are billions and billions of barrels of heavy oil in the world today and we are discovering more reserves of heavy oil every year.

2. In many parts of the world, light oil production is declining. Heavy oil conversion, therefore, becomes increasingly important to maintain economic viability of these regions.
3. Further tightening of environmental regulations in Europe and North America has provided incentives to refiners to further reduce sulfur and aromatic levels of the finished products.
4. Additional stringent requirements on the disposal of refinery residues such as coke or residua encourage *minimum waste* refinery strategies.
5. The market for heavier fuels is decreasing while that for middle distillate is increasing at a rapid rate.

During recent years, the average quality of crude oil has become worse because of a progressive decrease in API gravity (i.e., increase in density) and a rise in sulfur content. As the reserves of lighter crude oils continue to decline, there is an increasing need to develop options to upgrade the abundant supply of known heavy oil reserves into more value-added products.

2.0 FUNDAMENTALS OF UPGRADING PROCESSES

Heavy oils are generally characterized by low API gravity (high density) and high viscosity, high initial boiling point, high carbon residue, high nitrogen content, high sulfur content, and high metals content (Chapters 2 and 3). In addition, the hydrogen-to-carbon ratios of heavy oils are lower than desirable.

2.1 Upgrading Processes

Technologies for upgrading heavy crude oils such as heavy oil, bitumen, and residua can be broadly divided into *carbon rejection* and *hydrogen addition* processes.

Carbon rejection redistributes hydrogen among the various components, resulting in fractions with increased H/C atomic ratios and fractions with lower H/C atomic ratios. On the other hand, hydrogen addition processes involve reacting heavy crude oils with an external source of hydrogen and result in an overall increase in H/C ratio (Stanislaus and Cooper, 1994). Within these broad ranges, all upgrading technologies can be subdivided as follows:

1. *Carbon rejection*: visbreaking, steam cracking, fluid catalytic cracking, coking, flash pyrolysis.
2. *Hydrogen addition*: catalytic hydroconversion (hydrocracking) using active

hydrodesulfurization catalysts, fixed bed catalytic hydroconversion, ebullated catalytic bed hydroconversion, thermal slurry hydroconversion (hydrocracking), hydrovisbreaking, hydropyrolysis, donor solvent processes, supercritical water upgrading.

3. *Separation processes*: distillation, deasphalting, supercritical extraction.

In the following discussion, the term heavy crude oil will include heavy oil, tar sands bitumen, and residual oils (bottom of the barrel, vacuum bottoms).

2.1.1 Carbon Rejection

All upgrading technologies that chemically alter the crude oil rely on thermal treatment to crack large molecules to smaller fragments. At low severity, the main reactions involve cracking of long-chain hydrocarbons into lower molecular weight paraffins and olefins and the splitting off or cracking of aliphatic side chains attached to aromatic rings. This results in reduced viscosity and an increase in aromaticity but little conversion of high molecular weight material. As the severity increases, carbon-sulfur and carbon-nitrogen bonds split with elimination of hydrogen sulfide (H_2S) and ammonia (NH_3), and naphthenes dehydrogenate to produce aromatics. Polymerization also occurs, probably via radical mechanisms, resulting in formation of very high molecular weight coke product with a low atomic H/C ratio.

In the absence of a source of hydrogen, the end result is a range of hydrocarbons from methane (CH_4) to heavy gas oil plus up to ca. 35% coke. The yields and composition of the cracked products will depend on the reaction conditions and the presence or absence of cracking catalysts (Chapter 5). High temperatures and short residence times favor more gas and coke, while low temperatures and long residence time tend to produce less gas.

Although not strictly a refining process insofar as it is a nonthermal process, *solvent deasphalting* is, in fact, a carbon rejection process that is used during petroleum refining. A number of solvent deasphalting processes have been developed for removal of asphaltenes and asphaltic materials (asphaltenes plus resins) from heavy crude oils based on the use of light solvents such as propane, butane, and pentane to precipitate asphaltenes.

In the M.W. Kellog ROSE process (Low et al., 1995, Northrup and Sloan, 1996), the heavy crude oil is fed to a separator vessel and mixed with a stream of solvent at subcritical temperatures. The light solvent causes a phase separation and the denser asphaltene phase is removed from the bottom and stripped off any remaining solvent. The deasphaltened oil-solvent mixture exits from the top of the separator and is then heated to supercritical temperatures and passed into the deasphaltened oil separator. The deasphaltened oil is no longer soluble at this temperature and is removed from the bottom of the separator. The solvent exits from the top and is then cooled and recycled to the asphaltene separator. Because supercritical solvent recovery is lower in cost than flash vaporization, high solvent-to-feed ratios can be used, leading to maximum deasphaltened oil yield and quality.

Other solvent deasphalting technologies include IFP's Solvahl process (Peries et al., 1995), Foster Wheeler's LEDA process, Cosmo Oil's MDS process, and UOP's DEMEX process.

2.1.2 Hydrogen Addition

When thermal cracking occurs in the presence of hydrogen and a catalyst, reaction pathways change significantly. Thermal cracking still occurs, but hydrogenation and hydrogenolysis occur in parallel, changing the chemical nature of the molecules being cracked. Sulfur and nitrogen are removed from heterocyclic systems producing hydrogen sulfide (H_2S) and ammonia (NH_3) and formation of carbon-hydrogen bonds. The resultant aliphatic chains can then crack to produce light hydrocarbons such as methane, ethane, and propane. Hydrogenation of aromatics can occur at low temperatures with the formation of naphthenes; however, hydrogenation of aromatics is inhibited by sulfur and nitrogen compounds present in the feed (Stanislaus, 1994). Hydrogen can also cap radicals and terminate polymerization reactions, thereby reducing or eliminating coke formation. The catalysts normally employed are metal sulfide based and greatly accelerate hydrodesulfurization reactions leading to low-sulfur products. They probably do not directly catalyze cracking to any great extent. However, catalysts can be designed to accelerate the cracking reaction, although they are easily poisoned by metals normally present in heavy crude oils (Chapter 3). Although metal sulfides do catalyze hydrogenation of aromatics, because this reaction is reversible and very exothermic, temperatures normally employed to achieve high conversion of material boiling higher than 525°C (975°F) tend to favor the reverse reaction (dehydrogenation to aromatics). Unless operated at high hydrogen pressure and low liquid hourly space velocity (in order to reduce the temperature and still enable high conversions), most upgrading processes can achieve only low to moderate levels of aromatic saturation.

Yields of methane to pentane hydrocarbons can reach 10% w/w of feed at the highest severity (>90% residuum conversion). Because each mole of gas consumes approximately one mole of hydrogen, overall hydrogen consumption can reach 3% w/w of feed (approximately 2000 ft^3/bbl).

2.2 Chemistry of Upgrading

Heavy crude oils are composed chemically of a very broad range of molecules differing widely in molecular weight and chemical properties (Chapter 1). In addition, crude oils from different formations and locations have different compositions (Chapter 1). Because of the large number of components, it is not feasible to define heavy crude oils in terms of individual molecular components. Instead, various separation procedures are used to break down the feed into a number of smaller fractions. One such technique involves separation into solubility classes using solvents of varying polarity and further separation using column chromatography. These fractions can then be further characterized in terms of an *average structure* by techniques such as nuclear magnetic resonance.

Broadly speaking, heavy crude oils consist of paraffins, cycloparaffins (naphthenes), and aromatics of various ring sizes and degrees of aliphatic chain substitution, polar components, sulfur- and nitrogen-containing heterocyclic systems, and the like. The molecular weights range upward to many thousands (Chapters 1 and 4) and the boiling points reach 700°C (1290°F) or more.

Most crude oils are believed to be colloidal systems with micelles of high molecular weight polar components (asphaltenes) stabilized by components of intermediate polarity (resins). The asphaltene components contain most of the metals (vanadium, nickel, and iron) (Chapter 3) and form complexes with polydentate nitrogen and sulfur ligands such as porphyrins.

The main aim of upgrading is to decrease the molecular weight of large molecules to produce components with boiling points and H/C atomic ratios suitable for liquid fuels. At the same time, contaminants such as sulfur, nitrogen, and metals must be removed and aromatics saturated.

Ideally, different components require different processes and processing conditions to achieve the desired properties. For instance, hydrogenation of aromatics is best carried out at moderate temperatures with metal catalysts in the absence of sulfur and nitrogen compounds, while removal of sulfur and nitrogen uses metal sulfide–based hydrodesulfurization catalysts that need sulfur and operate at higher temperatures. The ideal process would involve separation of the crude oil into chemically different components and their treatment by the most suitable process. However, economic constraints preclude this approach. Normally, the only separation carried out is distillation to remove light fractions or solvent deasphalting to eliminate asphaltenes.

The objective of an upgrading process is to break chemical bonds in order to make lighter material (Chapter 5). The major bonds of interest during primary upgrading are carbon-carbon, carbon-sulfur, and carbon-hydrogen bonds. The typical means of doing this is by thermal reactions. The typical bond dissociation energies shown below (Benson, 1976, Chemical Rubber Co., Handbook of Chemistry and Physics, 1996) provide an estimate of difficulties in their breakage.

Bond dissociation energies, kcal/mole:

C–C (aliphatic)	85-90
C–H (n-alkanes)	95-105
C–H (aromatic)	110
C–SR dimethyl sulfide)	74
C–SH	75
C–O (methoxy)	82
C–N	84

Bond dissociation energies vary considerably, depending on factors such as the number and character of any substituents, chain branching, and the presence of double bonds. The resonance stabilization in aromatic compounds provide an extra 5–6 kcal/mole for carbon-carbon bonds. This stabilization of aromatic bonds render them unbreakable at process temperatures (<350°C; <660°F) until aromatic characteristics are destroyed by hydrogenation. For significant carbon-carbon bond breakage, a minimum temperature of approximately 420°C (790°F) is needed (Gray, 1994), especially without a catalyst., although $carbon_{aromatic}$-$carbon_{aliphatic}$ bond scission has been reported at temperatures as low as 320°C (610°F) for 4-(1-naphthylmethyl)bibenzyl in the presence of a catalyst (Farcasiu, 1991).

Catalysts such as zeolites can promote breaking of carbon-carbon bonds at hydrogenation temperatures below 400°C (750°F). However, zeolites are fouled rapidly in primary upgrading. The catalyst for carbon-carbon bond breakage relies on carbonium ion intermediates and is promoted by the acid sites on the catalyst surface. Superacids can break carbon-carbon bonds at a relatively low temperatures (25°C; 77°F) (Gray, 1994) but are rapidly deactivated by basic compounds such as nitrogen compounds.

The upgrading of heavy oil and residues is the result of free radical chain reactions. Free radicals are highly reactive intermediates that have an unpaired electron. The stability of alkyl free radicals is in the order:

tertiary > secondary > primary

Thus, the *t*-butyl radical (a tertiary radical) is energetically more favored than the ethyl radical. An example of a free radical reaction pathway is as follows:

1. *Initiation*:

$$M \rightarrow 2R \qquad (1)$$

M = mother molecule, reaction constant = k_1

2. *Propagation* (chain reaction):

$$R. + M \rightarrow RH + M. \qquad (2)$$

reaction constant = k_2

$$M \rightarrow R. + \text{olefins}$$

ß-Scission reaction, reaction constant, k_3

3. *Termination*:

$$R. + R. \rightarrow \text{products}$$

reaction constant, k_4

Free radical reactions are influenced by the hydrogen pressure. At pressures below 0.1 MPa, the formation of ethylene and propylene is favored. Indeed, these hydrocarbons are formed from light hydrocarbons, naphtha, and gas oil by thermal cracking in the temperature range 700–800°C (1290–1470°F).

At elevated pressures (>7 MPa), the reactions followed in low-pressure cases do not generally proceed. Under elevated pressure, hydrogen addition reactions become more favorable. Beta-scission reactions are less significant under elevated pressure. At elevated pressures, rather than multistage cracking via olefin formation, free radicals are stabilized in a single step without formation of olefins. In the intermediate pressure range of 3–7 MPa, a complicated two-step mechanism has been proposed (Gray, 1994).

The thiols (R–SH), aliphatic sulfides (R–S–R', thioethers) and disulfides (R–S–S–R') are very reactive under thermal conditions and can range as high as 50% of total sulfur in heavy oil, bitumen, and asphalt (Gray, 1994). Thermal reactions of these types of sulfur are favorable because carbon-sulfur bonds are weaker than other carbon-carbon bonds. For example:

$$R\text{–}S\text{–}R' + 2H_2 \rightarrow RH + R'H + H_2S$$

Thermal processing of these feedstocks at temperatures as low as 250°C (480°F) is accompanied by the formation of hydrogen sulfide. These thermal reactions of sulfur compounds could account for 30–50% of desulfurization without any catalysts at all. The thiophenic sulfur compounds are thermally stable, do not generally break down during primary upgrading, and require catalytic desulfurization. The action of catalysts provides the following two parallel reaction paths. The first reaction is referred to as hydrogenolysis (bond breakage by hydrogen). The second pathway is simple hydrogenation (hydrogen addition). In general, nickel-molybdenum (Ni/Mo) or cobalt-molybdenum (Co/Mo) catalysts are selective for sulfur heterocyclic compounds in preference to other aromatic hydrocarbons.

3.0 UPGRADING TECHNOLOGIES

There are many upgrading technologies from different licensers and having a wide range of installed capacity (Table 6-1).

3.1 Hydrogen Addition Processes

These all involve contact of a heavy crude oil with hydrogen at a pressure above 1000 psi and temperatures up to 470°C (880°F). The feedstock is thermally cracked and hydrogenated to yield products with increased H/C atomic ratio, reduced sulfur and nitrogen content, and boiling points suitable for refining to various liquid fuels. The processes can be divided into those employing high-activity hydrodesulfurization catalysts based on metals such as cobalt (Co), molybdenum (Mo), and nickel (Ni), which produce low-sulfur products, and those using less catalytically active additives or very low concentrations of a more active catalyst designed for coke inhibition and demetallization, which produce higher sulfur products requiring more extensive hydrotreatment.

Although the term *hydrocracking* is often applied to these processes, the term *hydroconversion* (or *hydroprocessing*) will be used here to avoid confusion with conventional gas oil hydrocracking.

The main features of all hydroconversion processes are similar. Heavy feedstock is preheated, mixed with hydrogen at pressure, and passed into a reactor kept at reaction temperature. Sometimes some or all of the hydrogen is added directly to the reactor. The residence time of liquid in the reactor can typically range from 1–10 hours. The hydroprocessed products then pass into a series of two or more vapor/liquid separators.

Table 6-1. Summary of upgrading processes.

Process	Licensor	Installed capacity, bbl/d	Comments	Reference
1. **HYDROGEN ADDITION** 1.1 **Catalytic Hydroprocessing Using HDS Catalysts** 1.1.1 **Fixed bed catalyst**				
RDS and VRDS Hydrotreating	Chevron	807,000 (1994), (3 VRDS units)	Graded catalyst system in one reactor. OCR (onstream catalyst replacemenr) option	Reynolds 1989, 1992
Unicracking/HDS	Unocal /UOP	430,000 (1982)	Multistage catalyst beds	Heckle 1996
Residfining	Exxon	160,000	One or more catalyst stages. Optional filter/guard bed	(1) p. 135
Hyvahl-F and S catalytic resid Hydrotreating	Asvahl/IFP	28,000	F—Two catalyst stages (i) conversion and demetalization, (ii) HDS and HDN S—includes guard bed	(1) p. 130, (7) p .59 Peries 1995
RCD Unibon	UOP	7,700 (1991)	One or two stages	(7) p. 68
(i) ABC/SDA (ii)ABC/HDS	Chiyoda	none	(i) Cracking and hydrodemetalization followed by solvent deasphaltening; (ii) Cracking and hydrodemetalization followed by HDS	Kakeuchi 1982
Residue Hydroconversion Process	Shell	79,000	Multibed reactors—demetallization followed by HDS and conversion.	(7) p 75
HYCAR Process	Showa Shell Sekiyu KK.	none	Visbreaking followed by two stage hydroporcessing	(7) p. 57
R-HYC Process	Idemitsu Kosan Co. Ltd.	none	Hydrocracking/HDS	(7) p. 73, Yamamoto 1991
RHC Process	Cosmo Oil Co. Ltd.	30,000 demo	Resid HDS combined with VGO HDS	(7) p. 72, Yamazaki 1988

Table 6-1. Summary of upgrading processes *(continued).*

Process	Licensor	Installed capacity, bbl/d	Comments	Reference
1.1.2 **Ebullated bed catalyst**				
H-Oil	HRI/Texaco	6 plants (1994)	External ebullating pump	Colyar 1988
LC-Fining	ABB Lummus Crest/ Oxy R&D/Amoco	6 plants (1994)	Internal ebullating pump	Reich 1993
1.2 **Thermal Slurry Hydroprocessing**				
CANMET Hydrocracking	Petro Canada/ SNC-Lavalin	5000	Iron-based additive; vapor phase HDS option	Benham 1995; Pruden 1993
VEBA Combi-cracking	VEBA	3500 (modified coal liquefication plant)	Iron-based additive; vapor phase HDS option	Niemann 1988; Wenzel 1993
Aurobon Process	UOP	None	Vanadium-based additive for demetallization	Anderson 1995
M-Coke	Imperial	None	Finely dispersed additive based on Mo or other metals	Bauman 1993
$(HC)_3$ Process	Alberta Research Council	None	Finely dispersed additive	Padamsy 1995
CANMET Emulsion Upgrading	CANMET, NRCan	None	Upgrades feed with high water and mineral matter content.	Patmore 1993
HDH Process	PDVSA/INTEVEP	15,000 (planned, 1995)	Adds clay minerals up to 10% of feed	Marzin 1995
HFC Process	NRIPR & Niigata Engineering Co. Ltd.	none	2–5% of spent HDS catalyst	Sakabe 1979
MRH Process	Idemitsu Kosan Co. Ltd. and M.W. Kellogg Co.	none	Coke deposits on "catalyst" and is withdrawn and burned off in regenerator	Sue 1989
SOC (Super Oil Cracking) Process	Asahi Chemical Industry Co., Nippon Mining Co. and Chiyoda Coorp.	3,500 demo	Uses finely dispersed additive	Seko 1989

Table 6-1. Summary of upgrading processes *(continued)*.

Process	Licensor	Installed capacity, bbl/d	Comments	Reference
1.3 **Noncatalytic — Hydrovisbreaking**				
Tervahl - H hydrovisbreaking Process	Asvahl/IFP	~600 demo	Use soaker visbreaker drum	(7) p. 26 Le Page 1987
1.4 **Hydropyrolysis**				
Hydropyrolysis	Univ. of Utah	None	Short residence time, high temperature.	Bunger 1988
1.5 **Donor Solvent Process**				
Donor refined Bitumen Process	Gulf	None		Logan 1985
1.6 **Supercritical Water Upgrading**				
Allied Process for Upgrading Heavy Crudes	Allied Corporation	None	Supercritical water	Marthy 1985, 1984; Patel 1988
2. **THERMAL REARRANGEMENT AND CARBON REJECTION** 2.1 **Visbreaking**				
Coil Visbreaking	ABB Lummus Crest Inc and others	~25 units (1992)		Singh 1986; (7) p. 27
Shell Soaker Visbreaking	ABB Lummus Crest Inc	>70 units (1994)		Singh 1986; (7) p. 27
Tervahl T Process	Asvahl	~350,000 (16.8 MMtpy) 1994		(7) p. 25
2.2 **Steam Cracking**				
Residue thermal cracking (HSC, high conversion soaker cracking)	Toyo Eng. Corp. & Mitsui Chem. Ltd.	14,000	Thermal cracking with simultaneous steam striping.	(7) p. 19; Washimi 1989
Eureka steam cracking	Kureha	~20,000	Thermal cracking with simultaneous steam striping.	(6) p. 161; Aiba 1981; Ruzhuo

Table 6-1. Summary of upgrading processes *(continued)*.

Process	Licensor	Installed capacity, bbl/d	Comments	Reference
ET-II Process	Fuji Oil Co. Ltd. and Chiyoda Corp.	none	Similar to Eureka steam cracking but lower costs are claimed	(7) p. 9; Watari 1987
2.3 **Fluid Catalytic Cracking**				
S & W FCC Process	Stone & Webster Eng. Corp/Total petroleum	300,000		(7) p. 43
Shell FCC Process	Shell	30 units		(7) p. 41; Khouw 1990
Flexicracking IIIR	Exxon	~17 units		Ladwig 1994
Fluid Catalytic cracking (FCC)	Texaco/ABB Lummus Crest,			(1) p. 111
RCC Fluid catalytic cracking	UOP	225,000		(1) p. 117; Myers 1994
HOC Process - Residue FCC	Kellogg	182,000 (1993)	Based on Orthoflow FCC reactor	(7) p. 33
R2R Resid FCC	IFP	334,000	Two-stage regenerator	(7) p. 37
Asphalt Residual Treating (ART) process	Engelhard	55,000 (1988)		Dennis 1988
MSCC	BARCO	55,000	Very short contact time; feed injected perpendicular to falling catalyst curtain	Kauff 1996
2.4 **Coking**				
Flexicoking	Exxon	160,000	Coke is gasified with air/steam	(1) p. 96; (7) p. 13; Guedez 1985
Fluid coking	Exxon	350,000	Coke is partially burned and hot coke recycled	(7) p. 15; Parish 1996
Delayed coking	ABB Lummus Crest; Kellogg; UOP: KoO Oil; Continetnal Oil	1,800,000		(7) p. 7 Feintuch 1986
Selective Yield Delayed Coking (SYDEC)	Foster Wheeler	58,000 tpd	Produces anode and needle coke	(1) p. 96

Table 6-1. Summary of upgrading processes *(continued)*.

Process	Licensor	Installed capacity, bbl/d	Comments	Reference
Lurgi Ruhrgas (LR) Coking process	Lurgi	200 tpd demo	Mechanical mixing of coke and vapor, short residence time	Weiss 1988
Dynacracking	HRI	None	Part of feed is gasified to H_2-containing gas which upgrades remaining feed	Rakow 1977
Cherry-P Process	Osaka Gas Co. Ltd.	36 t/d pilot	Cracks mixture of heavy crude and coal. Produces distillate and metalurgical coke binder pitch	(7) p. 5
FTC Process	Fuji Oil Co. Ltd./Fuji Standard Research Inc.	none	Coke is gasified.	(7) p. 17
KKI Process	Kobe Steel Ltd., Koa Oil Co. Ltd., Idemitsu Kosan Co. Ltd.	30 pilot	Cracking using fluidized bed of iron ore. Produces distillate and sponge iron	(7) p. 23
HOT (Heavy Oil Treating) Process	Nippon Mining co, Ltd., Hitachi Ltd., Kashima Oil Co Ltd. and RAROP	250 demo	Recirculating bed of iron oxide. In cracker $FeO + H_2O \rightarrow Fe_3O_4 + H_2$. In regenerator $Fe_3O_4 + C + O_2 \rightarrow FeO$	(7) p. 35
2.5 **Flash Pyrolysis**				
Flash Pyrolysis	NRCan	None	Temperature > 600°C, residence time <1 s. Produces high yields of ethene and propene	U.S. Patent 5,370,789
3. **SEPARATION PROCESSES** 3.1 **Solvent Deasphalting**				
ROSE Deasphalting	Kellogg (was Kerr-McGee)	221,000 (1994)	Supercritical solvent recovery	Northrup 1996 Low 1995

Table 6-1. Summary of upgrading processes *(continued)*.

Process	Licensor	Installed capacity, bbl/d	Comments	Reference
Solvahl Solvent Deasphalting process	Asvahl	32,000 tpy	Heavy solvent, low solvent ratio and supercritcal solvent recovery	(1) p. 100; (7) p. 99; Peries 1995
Low Energy Deasphalting (LEDA) process	Foster Wheeler	250,000 (42 units) designed	Liquid hydrocarbon solvent/rorating disk contactor	(7) p. 91
DEMEX	UOP	9 units	Supercritical solvent recovery	(7) p. 93
MDS	Cosmo Oil Co. Ltd.	None, (1 plant designed)	Supercritical solvent recovery; higher MW solvent	(7) p. 95

Typically, a hot high-pressure separator removes heavy liquid containing residuum and the vapor passes to a cold high-pressure separator to disengage gases from distillate product. Intermediate separators can be employed to reduce the temperature and pressure in stages. In some processes, a vapor phase hydrogenation unit is used to treat the vapor further before passing into the cold separator. Gas from the cold separator is then sent to a scrubber to remove hydrogen sulfide (H_2S), and ammonia (NH_3), and light hydrocarbons (which can be used as fuel gas), and the hydrogen gas is then recycled to the reactor. Fresh hydrogen, usually produced by steam reforming of methane, is added to make up for the hydrogen consumed.

Catalytic hydroprocessing using hydrodesulfurization catalysts employs an active hydrodesulfurization catalyst such as cobalt-molybdenum on alumina to remove most of the sulfur and some of the nitrogen present in the feedstock. At the same time, thermal cracking and hydrogenation of hydrocarbons cause conversion of high-boiling fractions to distillates. When operated at mild conditions, the process is usually referred to as hydrotreating and can be used to produce low-sulfur feedstock with a low Conradson carbon residue for a downstream fluid catalytic cracking unit.

A disadvantage of hydrodesulfurization catalysts is their rapid deactivation by metals and coke deposition on the active sites, necessitating continual catalyst replacement or use of a renewable catalyst guard bed. The processes can be divided according to the type of reactor used into fixed bed and ebullated bed reactors.

Fixed bed processes involve a fixed bed of catalyst, usually extruded into short cylindrical sections, packed into a tubular reactor through which heavy crude oil and hydrogen passes, usually in the downflow mode. Fixed bed hydroprocessing represents the commonest upgrading option currently in operation (Table 6-1).

Because heavy crude oils contain metals and constituents capable of forming high Conradson carbon residues (Chapters 1, 2, and 3), these processes suffer from catalyst deactivation due to the deposition of metals and coke on the catalyst surface. This limits the conversion or throughput that can be sustained. Strategies employed to deal with catalyst deactivation include demetallization guard beds and on-stream catalyst replacement techniques (such as Chevron's OCR technology). Residua conversions are normally limited to about 50% in once-through operation due to the high cost of catalyst replacement. High conversions can be attained by reducing the space velocity or recycling unreacted residuum. Sulfur removal can range up to 90%.

Most of the licensed processes have two or more stages. In the first, catalyst and conditions are chosen to crack and demetallize the feed. The second stage involves a hydrodesulfurization catalyst to further hydrogenate the feed and to remove sulfur and nitrogen and saturate some of the aromatics.

Ebullated bed processes involve suspending the catalyst in the feed by means of a feed recirculation pump, which pumps feed upward at sufficient speed to expand the catalyst bed at approximately 35% above the settled level. The feed returns to the recirculation pump via a central downcomer tube. This results in a back-mixed isothermal catalytic reactor where the liquid is thoroughly mixed with hydrogen gas and catalyst.

Two ebullated bed processes are currently being licensed: H-Oil licensed by HRI and Texaco (Colyar et al., 1988) and LC-Fining licensed by ABB Lummus Crest, Oxy R&D Co., and Amoco (Reich et al., 1993). Both processes use essentially the same technology but differ in some details such as the design and location of the ebullating pump. H-Oil uses an external pump, while LC-Fining relies on an internal ebullating pump. Both allow on-stream catalyst removal and replacement to maintain catalyst activity. Six H-Oil type units and six LC-Finers have been installed.

Residuum and sulfur conversions in the 90% range have been claimed for ebullated bed processes, although commercial units typically operate at lower severity to avoid excessive coke laydown and fouling.

Thermal slurry hydroconversion processes overcome some of the drawbacks of high catalyst cost and mechanical complexity of fixed and ebullated bed technologies (Del Bianco et al., 1993, 1995).

Slurry hydroconversion technology involves contact of heavy crude oils with hydrogen under pressure, usually in the presence of a mildly catalytic coke-inhibiting additive, at temperatures up to approximately 470°C (880°F). These additives can consist of low-cost iron-based materials or much lower concentrations of a finely divided molybdenum compound and are present as slurries of fine particles in the feed. At reaction conditions in the presence of hydrogen and sulfur-containing feeds, the active species are iron and molybdenum sulfides ($Fe_{1-x}S$ and MoS_2). Because iron-based catalysts are not very active during hydrodesulfurization and molybdenum-based catalysts are present in very low concentrations, sulfur removal is limited to about 50–60% of the original sulfur. In contrast, conversions in excess of 90% w/w of the 525°C (975°F) residuum can be achieved.

Much work has been carried out recently on the preparation and catalytic activity of finely dispersed (colloidal) metal sulfides for coal liquefaction and heavy oil upgrading. A number of processes have been developed and differ mainly in the type of additive used and the operating strategy employed.

CANMET Hydrocracking (Pruden et al., 1993), which is operating commercially, employs an iron sulfate-based additive and operates at medium pressure (<2000 psi hydrogen partial pressure). In order to achieve high throughputs using a single reactor, antifoam agents are used to maintain high liquid holdup. Recent improvements have allowed >90% residuum conversion using feedstocks with a high coking tendency by recycling residuum and heavy product fractions (Benham, 1996). One version of the CANMET Hydrocracking process, U-CAN, involves integration with vapor phase catalytic hydrogenation of the hot separator overheads (Pruden et al., 1993). CANMET Hydrocracking can accept a wide range of feedstocks, including bitumen and heavy oil with a high metals and asphaltene content. A 5000 bbl/day plant is currently in operation at Petro Canada's Montreal refinery.

Veba Combi-cracking (Niemann et al., 1988; Wenzel and Kretsmar, 1993) employs an iron-based additive but relies on use of higher pressures (up to 3500 psi) to control coke formation and reactor fouling. Hydrodesulfurization of the distillate product is achieved by passing the vapor from the hot separator to a gas phase hydrogenation unit. The technology has been demonstrated commercially in a 3500 bbl/day plant in Bottrup, Germany. More recently, INTEVEP (Marzin et al., 1995) has proposed a similar process called HDH, which is based on a catalyst prepared from a naturally occurring iron-containing ore.

The Aurabon Process developed by UOP (Anderson et al., 1995) is designed for removing asphaltenes and metals from feeds with high concentrations of these components to allow further upgrading by catalytic hydroprocessing. Hydrogen and heavy crude oil containing 2–6% finely divided vanadium sulfide (VS_2) and a solids dispersant are fed to a catalytic slurry reactor. Metals and coke from the feed presumably deposit on the particles of fluidized vanadium sulfide. The effluent stream from the reactor goes to a series of separators that yield four streams: (1) a gas stream, (2) a liquid stream, (3) a product stream, and (4) a drag stream, which contains most of the metal-containing solids. The gas stream and liquid stream are recycled. The product stream is low in asphaltenes and metals, whereas the drag stream is high in metals and solids and contains few unconverted hydrocarbons. The quality of the product stream can be controlled by varying conditions in the separators. No commercial plants have been built.

The use of ultrafine highly dispersed molybdenum or other metal-based catalysts has been studied by various groups. The catalysts can be produced by adding a catalyst precursor to the feed either in an oil-soluble form such as molybdenum naphthenate or as a very fine emulsion of an aqueous solution of a soluble salt such as ammonium molybdate. The ultimate form present in the reactor at conditions is molybdenum sulfide (MoS_2) and the aim is to achieve particles in the nanometer (meter x 10^{-9}) range. Exxon's M-Coke or Microcat process is based on this principle (Bearden, 1981; Bauman, 1993).

More recently, the Alberta Research Council proposed a similar upgrading process called $(HC)^3$ based on finely divided metal sulfides (Padamsey, 1995). Other proposed processes include HFC, which employs finely divided spent hydrodesulfurization catalyst (Sakabe, 1979); MRH, in which coke deposits on an unspecified catalyst and is burned off (Sue, 1989); and SOC (Super Oil Cracking) (Seko and Ohtake, 1989), which employs a very low concentration of an active catalyst (probably molybdenum based).

The upgrading of emulsions of heavy oil and water has been demonstrated in the CANMET Emulsion Upgrading process (Patmore et al., 1993) employing a slurry catalyst. The process uses syngas and the carbon monoxide reacts with water in the feed via the water gas shift reaction to generate hydrogen, which then hydrogenates the feed. The process is capable of upgrading bitumen containing high concentrations of fine mineral matter in addition to water. The unconverted residuum could be used to generate the syngas required for the process.

Hydrovisbreaking, a noncatalytic process, is conducted under similar conditions to visbreaking and involves treatment with hydrogen under mild conditions. The main objective is reduction of viscosity, although some conversion to lighter products does occur. The presence of hydrogen leads to more stable products (lower flocculation threshold) than can be obtained with straight visbreaking, which means that higher conversions can be achieved, producing a lower viscosity product. The Tervahl-H process (Institut Francais du Petrole, IFP) (LePage et al., 1987) is based on this principle.

Hydropyrolysis is a very short residence time (approximately 1 second) process operating at 500–560°C (930–1040°F) and is under development by the

University of Utah (Bunger et al., 1988). Low coke and gas yields at moderate conversion levels are claimed. Because the cracking and hydrogenation reactions occur mostly in the vapor phase, the rates of bimolecular reactions responsible for coke formation are lower and the concentration of aryl radicals, which can add to aromatic ring systems, is reduced by reaction with hydrogen.

Donor solvent processes are those processes in which an intermediate hydrogen transfer agent or hydrogen donor is used instead of directly hydrogenating heavy crude oils with hydrogen gas. The donor solvent reacts with the crude oil with transfer of hydrogen to produce hydrogenated products. The spent donor solvent can then be rehydrogenated.

In the Donor-Refined Bitumen (DRB) Process developed by Gulf Canada, bitumen is first fed into an atmospheric flash separator. The bottom product is then passed to a vacuum flash separator. The overhead distillates of both separators are sent to a crude oil reblending facility. Then, the bottom product of the vacuum flash separator is mixed with the hydrogen donor recycle stream, heated to reaction temperature, and sent to a series of reactors. No catalyst is used.

The liquid phase is separated by an atmospheric fractionator into a naphtha fraction, an exhausted middle distillate donor stream, and an atmospheric residuum. The naphtha is joined with the distillates of the original flash separators. The middle distillate donor stream is rehydrogenated using conventional fixed bed hydrogenation technology and is recycled. The atmospheric residuum is vacuum fractionated into a vacuum gas oil stream and a residuum stream (called Gulf-DRB residuum). The gas oil is mixed with the original distillate streams and the naphtha stream to form an upgraded reconstituted crude oil.

Supercritical water upgrading is a process to upgrade heavy crude oils by heating under pressure with water (Marthy, 1984, 1985; Patel et al., 1988). It is claimed that hydrogen from the water is incorporated in the crude oil, based on deuterium tracer experiments.

3.2 Thermal Rearrangement and Carbon Rejection

Thermal treatment of heavy crude oils in the absence of hydrogen induces hydrocarbon molecules to crack into smaller fractions, resulting in production of distillates and gas. At the same time, polymerization occurs with the formation of high molecular weight products (coke). This is the basis for a number of upgrading processes such as visbreaking, fluid catalytic cracking, and coking. These differ in the severity employed. Visbreaking is at the low end of the severity scale and is operated to minimize coke production. Coking operates at high severity and produces high yields of coke together with gas and distillate products.

Hydrogen is redistributed to give liquid products with higher H/C atomic ratios and solids with very low H/C atomic ratios.

Visbreaking is one of many processes that have been developed based on simple thermal treatment of heavy crude oils (Singh, 1986; Singh et al., 1993; Fainberg et al., 1996). Mild thermal treatment (*visbreaking*) results in rearrangement of the feed to produce a product with a lower viscosity but results only in low residuum conversion and sulfur removal (desulfurization: 5–10% for the Tervahl-T process). The process operates at low pressure and has the advantage of low cost. Visbreakers have been installed in many refineries and fall into two categories: coil visbreakers and soaker visbreakers.

The *coil visbreaker* employs a coil reactor and operates at higher temperatures and shorter residence times than the soaker visbreaker, which uses a drum reactor. The feed is heated directly in the coil visbreaker reactor and can result in more sediment (coke) and higher olefins due to the higher temperature used.

The *soaker visbreaker* preheats the feed, which then passes into a soaking drum where most of the cracking occurs. The temperatures used in soaker visbreaking are approximately 410–440°C (770–825°F) with residence times of 8–32 minutes. Coil visbreakers operate at approximately 470–500°C (880–930°F) with residence times of 0.5–2 minutes (Singh, 1986). Shell soaker visbreaking (licensed by ABB Lummus Crest) is the more popular process, and over 70 units have been installed or are in design and construction, while only approximately 25 coil visbreakers have been installed. Asvahl has also licensed 17 Tervahl-T soaker visbreakers.

Visbreaking cracks aliphatic side chains and naphthenic rings. This results in increased aromaticity and decreased viscosity of the product.

Steam cracking involves treating heavy crude oil with superheated steam at temperatures of 440–460°C (770–860°F) for the HSC process (Washimi, 1989), 490–495°C (915–925°F) for the ET-II process (Watari et al., 1987), and up to 500°C (930°F) for the Eureka process (Aiba et al., 1981). The actual temperature used depends on the feedstock and is adjusted to minimize coke formation. The crude oil is cracked and the distillate products are stripped off by the steam. The residuum product can be used for fuel or as metallurgical coke. Higher conversions than obtained with visbreaking are reported, and the presence of steam and the high turbulence in the reactor are claimed to prevent coke deposition. The original Eureka process uses two batch reactors, while the newer ET II and HSC processes both employ continuous reactors.

Fluid catalytic cracking (FCC) traditionally cracks vacuum gas oils etc. to coke, gasoline, and lighter distillate fractions using a fluidized cracking catalyst (Sadeghbeigi, 1995; Magee and Mitchell, 1993). The coke deposits on the catalyst, which is separated from the products, the coke is burned off, and the catalyst is recycled. The combustion of coke supplies the heat for the cracking reactions.

Over the past 15 years, use of fluid catalytic cracking technology to convert residues has grown rapidly; total residue fluid catalytic cracking (RFCC) capacity is now about 2 million bbl/day in 48 units worldwide (Dickenson et al., 1996). Residue fluid catalytic cracking units are designed to crack heavy crude oils but, because cracking catalysts are poisoned by >25 ppm metals and feeds with >7% Conradson carbon residue produce excess heat, such heavy feeds must be pretreated to remove these undesirable components. This can be done by hydrotreating the feedstock using hydrodesulfurization catalytic processes. Recent advances in vanadium trapping and metal-tolerant catalysts (e.g., Dougan et al., 1994; Wolterman et al., 1996) and use of catalyst cooling (Myers et al., 1994) allow the use of feed with higher metals and Conradson carbon residues.

Other recent developments include the short contact time MSCC process, in which feed is injected perpendicular to a downflowing stream of catalyst. The short contact time combined with a low-volume reaction zone reduces secondary cracking and produces more gasoline and less coke compared with conventional fluid catalytic cracking. The lower coke yield allows feeds with higher Conradson carbon residues to be processed (Kauff et al., 1996).

The ART process is similar to residue fluid catalytic cracking processes but can handle heavy crude oils with Conradson carbon residues of at least 13.5% and a metals loading of 300 ppm (Green and Center, 1985). This process uses a proprietary fluidized catalyst on which coke and metals deposit. The coke is burned off and the catalyst recycled. Metals loading of up to 3% w/w can be achieved before the catalyst is regenerated by removal of metals.

Coking technology, which has been practiced since the late 1800s, takes heavy crude oils and thermally rearranges them to produce carbon and distillate. Hydrogen is transferred from high molecular weight components to low molecular weight components. Most of the metals remain with the coke, although the sulfur content of the liquid product is still high (Chapters 2 and 3). In order to remove sulfur and to saturate any unsaturated compounds produced during cracking, the products are hydrotreated to produce a synthetic crude oil for further refining.

In the *delayed coking* processes such as those licensed by ABB Lummus Crest and Kellogg, heavy crude oil is preheated to coking temperature and then passes into a coking drum, where the coking process is completed (Feintuch et al., 1986). Coke collects in the drum and the product exits as vapor which is then condensed. When the coke drum is full, hot feed is switched to a second drum, coke is removed from the first drum, and so on.

Fluid coking licensed by Exxon is similar to fluid catalytic cracking but does not employ a catalyst (Parish et al., 1996). Instead, heavy crude oil is fed to a fluidized bed of hot coke, where it cracks and coke deposits on the fluidized coke particles. The coke is withdrawn from the bottom of the reactor and fed to a fluidized bed combustor, where up to 30% of the coke is burned off to supply heat for the cracking step. The hot coke is then fed to the top of the fluidized coke bed in the cracking vessel. The typical range of products is 10–13 wt. % gas, 52–66 wt. % liquids and 24–35 wt. % coke (70–85% of the produced coke is burned for process heat). The products contain up to approximately 50% of the sulfur present in the feed and must be hydrotreated before further refining.

Flexicoking is a variant of Exxon's Fluid Coking process wherein part of the coke is gasified with air/steam instead of being combusted (Hydrocarbon Processing, 1996). Feed is cracked by hot coke recirculating through a heater that is heated by hot coke and gas from the gasifier, which in turn is fed by coke recirculated from the heater. Gas is sent from the heater to cleanup to produce low calorific fuel gas. A 1% coke purge removes 99% of the feed metals.

The LR *flash coking* process developed by Lurgi (Weiss et al., 1988) employs a mechanical radial mixer to contact hot granular coke with heavy crude oil. Because the vapor residence time is low (ca. 1 second), Lurgi claims that flash coking produces less coke and gas and more liquid than delayed or fluid coking. Short residence time also minimizes dehydrogenation reactions and the formation of cyclic and aromatic compounds, resulting in a liquid product with a higher H/C atomic ratio. Flash coking products contain 82% of the hydrogen originally present in the feed, compared with 73% for fluid coking and 64% for delayed coking.

Flash pyrolysis is carried out at very high temperatures (>600°C, 1110°F) and low residence time (<1 second) and converts heavy crude oils mainly to gas. The gases formed are very olefinic and could be used as chemical feedstocks. Natural Resources Canada has developed a process that produces ethylene and propylene in yields of 50 to 70% w/w of feed (U.S. Patent 5,370,789).

4.0 REFERENCES

Aiba, T., Kaji, H., Suzuki, T., and Wakamatsu, T. 1981. Chem. Eng. Progr. February: 37.

Anderson, R. F., Olson, R. K., Hutchings, L. E. and Penning, R. T. 1982. Proceedings. 2nd UNITAR International Conference on Heavy Crude and Tar Sand Vol. 3. Chap. XVII.

Bauman, R. F., Aldridge, C. L., Bearden, R., Jr., Mayer, F. X., Stuntz, G. F., Dowdle, L. D. and Fiffron, E. 1993. Preprints. Oil Sands–Our Petroleum

Future. Alberta Research Council, Edmonton, Alberta, Canada. p. 269.
Bearden, R., and Aldridge, C. L. 1981. Energy Progr. 1: 44.
Benham, N. K., and Pruden, B. B. 1996. National Meeting. National Petroleum Refiners Association. San Antonio, Texas, March 17–19. Paper AM-96-58.
Bunger, J. W., Tsai, C. H., and Devineni, P. 1988. Proceedings. 4th UNITAR/UNDP International Conference on Heavy Crude and Tar Sand, Edmonton, Alberta, Canada. 5: 169.
Colyar, J., Harris, E., and Popper, G. 1988. Proceedings. 4th UNITAR/UNDP International Conference on Heavy Crude and Tar Sand. Edmonton, Alberta, Canada. 5: 215.
Del Bianco, A., Panariti, N., Prandini, B., Beltrame, P. L., and Carniti, P. 1993. Fuel. 72: 81.
Del Bianco, A., Panariti, N., Beltrame, P. L., and Carniti, P. 1993. Fuel. 72: 75.
Del Bianco, A., Garuti, G., Pirovano, C., and Russo, R. 1995. Fuel. 74: 756.
Del Bianco, A., Panariti, N., and Marchionna, M. 1995. CHEMTECH. November: 35.
Dennis, G. E. 1988. Proceedings. 4th UNITAR/UNDP International Conference on Heavy Crude and Tar Sand. Edmonton, Alberta, Canada. 5: 89.
Dickenson, R. L., Schulman, B. L., Biasca, F. E., and Johnson, H. E. 1996. Annual Meeting. National Petroleum Refiners Association. San Antonio, Texas, March 17–19. Paper AM-96-57.
Dougan, T. J., Alkemade, U., Lakhanpal, B., and Boock, L. T., 1994. Annual Meeting. National Petroleum Refiners Association. March 20–22. San Antonio, Texas. Paper AM-94-46.
Fainberg, V., Podorozhansky, M., Hestroni, G., Brauch, R., and Kalchouk, H. 1996. Fuel Sci. Technol. Int. 14: 839.
Farcasiu, M., and Smith, C. 1991. Energy and Fuels. 5: 83.
Feintuch, H. M., Bonillo, J. A. and Godino, R. L., 1986. In Handbook of Petroleum Refining Processes. R. M. Meyers (Editor). McGraw-Hill, New York. Chapter 7.1.
Gray, M. R. 1994. Upgrading Petroleum Residues and Heavy Oils. Marcel Dekker Inc., New York.
Green, P. M., and Center, A. M. 1985. Proceedings. 3rd UNITAR/UNDP International Conference on Heavy Crude and Tar Sands. Long Beach, California. p. 1219.
Guedez, E. J., 1985. Proceedings. 3rd UNITAR/UNDP International Conference on Heavy Crude and Tar Sand. Long Beach, California. p. 1395.
Heckel, T. L., Anderle, C. J., and Reno, M. E., 1996. Annual Meeting. National Petroleum Refiners Association. San Antonio, Texas. Paper AM-96-61.
Hydrocarbon Process. 1996. November: 89.

Kauff, D. A., Bartholic, D. B., Steves, C. A., and Keim, M. R. Annual Meeting. National Petroleum Refiners Association. San Antonio, Texas. Paper AM-96-27.

Khouw, F. H. H. 1990. Spring Meeting. National Petroleum Refiners Association. Paper Am-90-42.

Ladwig, P. K., Bienstock, M. G., Citarella, V. A., and Draemel, D. C. 1994. Annual Meeting. National Petroleum Refiners Association. San Antonio, Texas. Paper AM-94-42

LePage, J. F., Morel, F., Trassard, A. M., and Bousquet, J. 1987. Preprints Div. Fuel Chem. 32: 470.

Logan, A., Simpson, P. L., Souhrada, F., and Woods, H. J. 1985. Proceedings. 3rd UNITAR/UNDP International Conference on Heavy Crude and Tar Sand, Long Beach, California. p. 1213.

Low, J. Y., Hood, R. L., and Lynch, K. Z. 1995. Preprints Div.Petrol. Chem. Am. Chem. Soc. 40: 780.

Magee, J. S., and Mitchell M. M., Jr. 1993. Fluid Catalytic Cracking: Science and Technology. Studies in Surface Science and Catalysis. Volume 76. Elsevier, Amsterdam, The Netherlands.

Marzin, R., Solari, B., and Duque, J. 1995. Proceedings. 6th UNITAR International Conference on Heavy Crude and Tar Sand. Houston, Texas. 2: 493.

Meyers, R.A. (Editor). 1986. Petroleum Refining Processes. McGraw-Hill, New York.

Murthy, A. K. S., Patel, K. M., and Bekker, A. Y. 1984. Process for production of light hydrocarbons by treatment of heavy hydrocarbons with water. U.S. Patent 4,446,012.

Murthy, A. K. S., Patel, K. M., and Bekker, A. Y. 1985. Proceedings. 3rd UNITAR/UNDP International Conference on Heavy Crude and Tar Sand. Long Beach, California. p. 1291.

Myers, D. N., Knapik, P. G., Lacijian, L. A., and Brandner, K. J. 1994. Annual Meeting. National Petroleum Refiners Association. San Antonio, Texas. Paper AM-94-26.

Niemann, K., Kretschmar, K., Rupp, M., and Merz, L. 1988. Proceedings. 4th UNITAR/UNDP International Conference on Heavy Crude and Tar Sand. Edmonton, Alberta, Canada. 5: 225.

Northrup, A. H., and Sloan, H. D. 1996. Annual Meeting. National Petroleum Refiners Association. Houston, Texas. Paper AM-96-55

Padamsey, R., Bailey, R. T., and Cyr, T. J. 1995. Proceedings. 6th UNITAR International Conference on Heavy Crude and Tar Sand. 2: 261.

Parish, M. R., Hammond, D. G., and Citarella, V. A. 1996. Hydrocarbon Technol. Int. Spring: p. 25.

Patel, K. M., Murthy, A. K. S., and Bekker, A. Y. 1988. Catalytic process for production of light hydrocarbons by treatment of heavy hydrocarbons with water. U.S. Patent 4,743,357.

Patmore, D. J., Hogan, C. H., and de Bruijn, T. J. W. 1993. Preprints. Oil Sands–Our Petroleum Future. Alberta Research Council, Edmonton, Alberta, Canada. p. 351.

Patmore, D. J., Hogan, C. H., and de Bruijn, T. J. W. 1993. Preprints. Oil Sands–Our Petroleum Future. Alberta Research Council, Edmonton, Alberta. p. 351.

Peries, J. P., Billon, A., Hennico, A., Morrison, E., and Morel, F. 1995. Proceedings. 6th International Conference on Heavy Crude and Tar Sand, Houston. 2: 229.

Pruden, B. B., Muir, G., and Skripek, M. 1993. Preprints. Oil Sands - Our Petroleum Future. Alberta Research Council, Edmonton, Alberta, Canada. p .277.

Rakow, M. S. 1977. Oil & Gas J. August 8.

Rakow, M. S. 1977. 83rd National AIChE Meeting. Houston, Texas. March 20–24.

Reich, A., Bishop, W., and Veljkovic, M. 1993. Preprints. Oil Sands–Our Petroleum Future. Alberta Research Council, Edmonton, Alberta, Canada. p. 216.

Reynolds, J. G. 1989. Annual Meeting. National Petroleum Refiners Association. San Francisco, California.

Reynolds, J. G. 1992. Annual Meeting. National Petroleum Refiners Association. New Orleans, Louisiana.

Ruzhuo, C. 1993. Preprints. Oil Sands–Our Petroleum Future. Alberta Research Council, Edmonton, Alberta, Canada. p. 287.

Sadeghbeigi, R. 1995. Fluid Catalytic Cracking Handbook. Gulf Publishing Co., Houston, Texas.

Sakabe, T. 1979. Hydrocarbon Process. 58(12): 109.

Seko, M., and Ohtake, N. 1989. AIChE Symposium Series No. 273. Vol. 85.

Singh, V. D. 1986. Erdöl und Kohle. 39: 19.

Singh, I. D., Koththiyal, V., Kapoor, M. P., Ramaswamy, V., and Aloopwan, M. K. S. 1993. Fuel. 72: 751.

Stanislaus, A., and Cooper, B. H. 1994. Catal. Rev. Sci. Eng. 36: 75.

Sue, H. 1989. Proceedings. 4th UNITAR/UNDP Conference on Heavy Oil and Tar Sands. 5: 117.

Takeuchi, C., Komatsu, S. and Kashiwara, H., 1982. Proceedings. 2nd UNITAR International Conference on Heavy Crude and Tar Sands. p. 1227.

Washimi, K. 1989. Hydrocarbon Process. 68(9): 69.
Washimi, K., and Hamamura, M. 1993. Preprints. Oil Sands–Our Petroleum Future. Alberta Research Council, Edmonton, Alberta, Canada. p. 283.
Watari, R., Shoji, Y., Ishikawa, T., Hirotani, H., and Takeuchi, T. 1987. Annual Meeting. National Petroleum Refiners Association. San Antonio, Texas. Paper AM-87-43
Weiss, H. J., Schmalfeld, J., and Solari, R. B. 1988. Proceedings. 4th UNITAR/UNDP International Conference on Heavy Crude and Tar Sand. Edmonton, Alberta, Canada. 5: 181.
Wenzel, F., and Kretsmar, K. 1993. Preprints. Oil Sands–Our Petroleum Future. Alberta Research Council, Edmonton, Alberta, Canada. p. 248.
Wolterman, G. M., Dodwell, G., and Lerner, B. 1996. Annual Meeting. National Petroleum Refiners Association. San Antonio, Texas. Paper AM-96-46
Yamamoto, T. 1991. Proceedings. 13th. World Petroleum Congress, Buenos Aires, Argentina.
Yamazaki, H., Tawara, K., and Tomino, T. 1988. Annual Meeting. National Petroleum Refiners Association. San Antonio, Texas.

Chapter 7

Hydrocracking: Reactions, Catalysts, and Processes

Geoffrey E. Dolbear
G. E. Dolbear & Associates
23050 Aspen Knoll Drive
Diamond Bar
California 91765

1.0 INTRODUCTION

Hydrocracking is a process for converting large molecules in crude oil into more valuable smaller molecules that can be used in transportation fuels. It finds its roots in fluid catalytic cracking and processes for hydrogenation of petroleum and coal tar. These processes were developed in the 1920s and 1930s. Modern hydrocracking traces its roots to the invention of zeolite catalysts in the 1960s.

Hydrocracking catalysts typically contain separate hydrogenation and cracking functions. Palladium sulfide and promoted group VI sulfides (nickel molybdenum or nickel tungsten) provide the hydrogenation function. These active compositions saturate aromatics in the feed, saturate olefins formed in the cracking, and protect the catalysts from poisoning by coke. Zeolites or amorphous silica-aluminas provide the cracking functions. The zeolites are usually type Y (faujasite), ion exchanged to replace sodium with hydrogen and make up 25 to 50% of the catalysts. Pentasils (silicalite or ZSM-5) may be included in dewaxing catalysts.

Basic nitrogen compounds, typically quinoline and its homologs, can poison the acid sites in these catalysts. This is especially a problem with zeolites. Hydrocracking processes using these catalysts will include a separate pretreater for hydrodenitrogenation (HDN) of the feed. Hydrodenitrogenation catalysts usually contain molybdenum sulfide promoted with nickel, with alumina as the high surface area support. Ammonia formed in the hydrodenitrogenation reaction is a less effective catalyst poison than quinoline and need not be removed from the mixture fed to the hydrocracking reactor.

Processes operate with high pressures of hydrogen, typically 1500 to 2500 psi, and temperatures are 290 to 370°C (550 to 700°F). Several process configurations are used, depending on the feed and the design criteria. All include provisions for addition of cold hydrogen at several points in the hydrocracking reactor to control reactor temperatures, since a great amount of heat is released by hydrogenation. Reactor internals provided for this function are complex mechanical devices.

Hydrocracking processes convert aromatic gas oils into high-quality gasoline, diesel, and turbine stocks. In doing this, they saturate aromatic rings, crack cycloparaffins (naphthenes) and paraffins, and saturate olefins formed during cracking. The organic chemistry of these steps is well known.

Hydrocracking processes are widely used (Table 7-1), especially when dealing with aromatic feedstocks such as those found in California. They are not as widely used as fluid catalytic cracking because the high hydrogen pressures used in hydrocracking lead to high capital costs that are hard to justify for less aromatic feeds.

Two quite different kinds of reactions take place simultaneously in hydrocrackers. Aromatic rings are saturated to cycloparaffins, while paraffinic structures are cracked to generate products with lower boiling ranges. This combination results in products with lower average molecular weights and lower overall aromaticity, an ideal combination for producing gasoline, diesel, and turbine fuel meeting today's more stringent emission specifications.

Table 7-1. Hydrocracking is a widely used process.

	United States	Western Europe	Asia/Pacific
Hydrocracking		580	531
Fluid Catalytic cracking	5284	1944	2209
Total refining capacity	15,354	12,152	14,802

Capacities, in 1000 bbl per day; Oil and Gas J. 1995. December: p. 43.
Western Europe includes Belgium, Denmark, France, Germany, Greece, Italy, Netherlands, Portugal, Spain, and UK.
Asia/Pacific includes Australia, Bangladesh, Brunei, China, Indonesia, Japan, Korea, Malaysia, Myanmar, New Zealand, Pakistan, Philippines, Singapore, Sri Lanka, Taiwan, and Thailand.

Catalysts for hydrocracking contain components for both the hydrogenation and cracking reactions. Hydrogenation activity is provided by palladium (Pd) or promoted molybdenum (Mo) or tungsten (W) sulfides. Cracking takes place on strong acid sites in zeolites or amorphous silica aluminas. Specialty catalysts including narrow pore zeolites are used in dewaxing lubricating (lube) oil stocks.

Basic nitrogen compounds such as quinoline can poison the acid sites. They are usually removed in a pretreating step, typically with a nickel/molybdenum sulfide catalyst that also removes sulfur.

2.0 CRACKING PROCESSES

Crude oil contains many thousands of different compounds. These range in molecular weight from a low of 16 for methane (CH_4) to more than 2000 (Boduszynski 1987, 1988; Speight, 1991). This broad range in molecular weights results in a boiling points that range from -160°C (-288°F) to nearly 1100°C (2000°F). Of course, most hydrocarbons decompose thermally at temperatures above about 650°F (340°C), so these higher boiling points cannot be measured directly and must be estimated from other measurements.

For a petroleum refinery, the highest value products are transportation fuels:

- Gasoline (b.p. < 220°C, < 425°F)
- Jet fuel (b.p. 175–290°C, 350°F–550°F)
- Diesel (175–370°C, 350°F–700°F)

Less than one third of a typical crude distills in these ranges. The heavier fuel fractions can be burned directly, but this is not an efficient use of the resource. Refiners use various thermal and catalytic processes to convert molecules in the heavier fractions to smaller molecules in fractions distilling at these lower temperatures. This efficiency translates into a strong economic advantage, leading to widespread use of conversion processes in refineries today.

Thermal (noncatalytic) cracking processes include coking and visbreaking. They are generally inexpensive but not very selective. Coke and visbroken fuel oil are low-value byproducts. Coke, in fact, is sold for about $5 a ton, the equivalent of oil at about $1 per barrel. Flexicoking, making fine coke as an intermediate that is gasified in a second step, is a more expensive process to build. Since it makes hydrogen instead of coke, it provides important economic benefits.

Catalytic processes include fluid catalytic cracking (FCC) and hydrocracking (HC). They are typically more expensive to build and operate. They also are limited in the amount of very heavy molecules they can handle, because the highly aromatic heavy molecules tend to produce large amounts of coke on the acidic catalysts. Both fluid catalytic cracking and hydrocracking make high yields of valuable liquids that can be blended directly into transportation fuels. For this reason, catalytic cracking processes have been described as violating the traditional dictum that *you can't make a silk purse out of a sow's ear*.

All of the thermal and catalytic processes are available in a broad variety of configurations from companies around the world. The detailed catalysts and process configurations vary among the developers, but the chemical principles involved are the same.

3.0 HISTORY

Catalytic cracking processes evolved in the 1930s from research on petroleum and coal liquids. The petroleum work came to fruition with the invention of acid cracking by Eugene Houdry. The coal liquid work, most notably in Germany, resulted in metal sulfide hydrogenation catalysts.

Eugene Houdry, in the 1930s, developed a catalytic cracking catalyst for petroleum that used solid acids as catalysts. His first catalysts were acid-treated clays, formed into beads. Clays are a family of crystalline aluminosilicate solids, and the acid treatment develops acidic sites by removing aluminum from the structure. The acid sites also catalyze the formation of coke, and Houdry developed a moving bed process that continuously removed the coked beads from the reactor for regeneration by oxidation with air (McEvoy, 1996).

Clays are natural compounds of silica and alumina, containing major amounts of the oxides of sodium, potassium, magnesium, calcium, and other alkali and alkaline earth metals. Iron and other transition metals are often found in natural clays, substituted for the aluminum cations. Oxides of virtually every metal are found as impurity deposits in clay minerals.

Clays are layered crystalline materials. They contain large amounts of water within and between the layers (Keller, 1985). Heating the clays above 100°C can drive out some or all of this water; at higher temperatures, the clay structures themselves can undergo complex solid-state reactions. Such behavior makes the chemistry of clays a fascinating field of study in its own right.

Typical clays include kaolinites, montmorillonites, and illites (Keller, 1985). They are found in most natural soils and in large, relatively pure deposits, from which they are mined for applications ranging from adsorbents to paper making.

Strong mineral acids react with clays to hydrolyze aluminum-oxygen bonds. The aluminum dissolves, leaving an altered structure. Protons in the acid also exchange with sodium and other alkali cations, leaving an acidic material. Houdry took advantage of such reactions in making the active acidic catalysts for his catalytic cracking process.

Later work at Exxon (then Esso) extended Houdry's work. The Esso workers built on the new technology of fluidized beds, using finely divided catalysts that act as a fluid when entrained in a flowing gas (Jahnig *et.al.*, 1984). Their fluid catalytic cracking process soon came to dominate, and by the mid-1960s very few bead catalyst units remained. Synthetic silica alumina catalysts were found to be more active, selective, and stable than the acid-leached clays.

The development of zeolite catalysts at Mobil in the 1960s took fluid catalytic cracking another step forward. The zeolite catalysts are far more active and, more important, far more selective than the amorphous catalysts. The result is today's highly versatile processes. Much has been written elsewhere about this development (Avidan, 1983).

Hydrogenation catalysts emerged from early coal conversion processes developed by Bergius. Work in the 1930s under Matthias Pier used metal sulfide catalysts to hydrocrack coal tars. Of most importance today is the family of molybdenum sulfide catalysts and the closely related tungsten sulfides. These are usually promoted with cobalt or nickel. The importance of this advance should not be underestimated, since it allowed catalysts to survive the potentially poisoning effects of sulfur in the petroleum and coal liquids.

Catalytic hydrocracking was a natural outcome of these two advances, using the hydrogenation activity of the metal sulfides and the cracking activity of silica alumina catalysts.

At about the same time that zeolites were being applied to fluid catalytic cracking at Mobil, Rowland Hansford at Unocal recognized the potential for zeolites in hydrocracking. His work led to the first modern hydrocrackers, with much higher activities and high selectivities. It should be noted that processes using noncrystalline silica alumina catalysts are still used today.

4.0 HYDROCRACKING

Hydrocracking processes convert gas oils to high yields of transportation fuels. The hydrogenation function ensures that aromatic levels are reduced compared with fluid catalytic cracking, and olefin levels are essentially zero. This makes hydrocracked diesel and turbine stocks far superior to fluid catalytic cracking products in such properties as cetane number, smoke point, and sulfur levels. The reduced aromatics and olefins mean that hydrocracked gasolines are lower in octane than fluid catalytic cracking gasolines. These stocks are, however, excellent blending stocks for reformulated gasolines.

A major economic advantage of hydrocracking comes from the phenomenon called *volume swell.* The volume density of aromatic liquids is higher than that of paraffins (Table 7-2). In the course of the hydrocracking process, hydrogen addition converts enough of the aromatics to paraffins to have an important effect on product density. There is only a little loss of material from the liquid fraction due to gas formation.

The net result is an increase (or swell) in volume, with the product often 20% or more larger than the feed. Since petroleum products are sold by volume, not mass, this is a distinct economic advantage for hydrocracking processes.

4.1 Hydrocracking Chemistry

Petroleum fractions contain hundreds or thousands of compounds, making the detailed chemistry exceedingly complex. To simplify the description of these systems, petroleum chemists usually distinguish two broad classes of reactions that take place in a hydrocracking system—hydrogenation and cracking.

Hydrogenation takes place on the metal sulfide sites. Hydrogen is added to aromatics, converting them to cycloparaffins. Olefins formed in cracking reactions are also hydrogenated.

Cracking takes place on strong acid sites, opening rings and cracking chains into smaller pieces; octyl naphthalene offers an example of these reactions

Table 7-2. Hydrogenation decreases density.

Compound	Density
Benzene	0.879
Cyclohexane	0.779
Naphthalene	1.145
Tetrahydronaphthalene	0.970
Decahydronaphthalene (cis)	0.895

(Figure 7-1). The first step in the reaction is saturation of one of the rings to form a substituted tetralin (tetrahydronaphthalene) intermediate. Saturation creates a paraffinic structure that is reactive to cracking with strong acid catalysts, whereas the original aromatic ring is quite inert. Once the tetralin is formed, acid-catalyzed cracking leads to ring opening. The product is a terminal olefin. Olefins are not stable in the vigorous hydrogenation environment and saturate to form paraffins.

The substituted naphthalene could also have taken a second route, with the alkyl side chain cracking (lower part of Figure 7-1). Of course, the naphthalene products could react further by hydrogenation and cracking to give single-ring products. This simplified network illustrates the potential complexity of the system.

Such cracking reactions were explored in classic studies in the 1940s and 1950s, and the mechanisms summarized in the classic work of Greensfelder, Voge and Good (Greensfelder et al., 1949).

5.0 HYDROGENATION CATALYSTS

The very best catalysts for saturating aromatic structures are found among formulations made with platinum and nickel. Unfortunately, both of these metals are sensitive to poisoning by sulfur. They cannot be used in petroleum hydrocracking because petroleum invariably contains sulfur compounds.

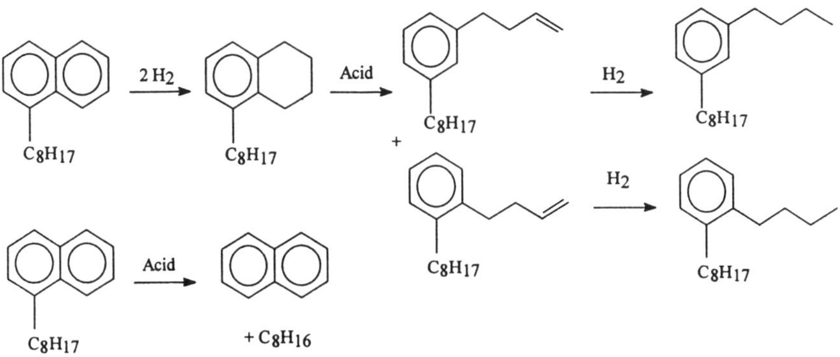

Figure 7-1. Octyl naphthalene and possible hydrocracking routes.

Modern hydrocracking catalysts rely on less active metal sulfides that do not suffer from this sensitivity. The most common among these are promoted molybdenum sulfides and palladium sulfide.

Molybdenum sulfide (MoS_2), usually supported on tailored alumina, is widely used in petroleum processes for hydrogenation reactions. It is a layered structure that can be made much more active by addition of cobalt or nickel (Topsoe *et al*, 1996). When promoted with cobalt sulfide (CoS), making what is called *cobalt-moly* catalysts, it is widely used in hydrodesulfurization (HDS) processes. The nickel sulfide (NiS)–promoted version is used for hydrodenitrogenation (HDN) as well as hydrodesulfurization The closely related tungsten compound (WS_2) is used in commercial hydrocracking catalysts. Other sulfides (iron sulfide, FeS, chromium sulfide, Cr_2S_3, and vanadium sulfide, V_2S_5) are also effective and used in some catalysts. A valuable alternative to the base metal sulfides is palladium sulfide (PdS). Although it is expensive, palladium sulfide forms the basis for several very active catalysts.

The choice of hydrogenation catalyst depends on what the catalyst designer wishes to accomplish. In catalysts to make gasoline, for instance, vigorous cracking is needed to convert a large fraction of the feed to the kinds of molecules that will make a good gasoline blending stock. For this vigorous cracking, a vigorous hydrogenation component is needed. Since palladium is the most active catalyst for this, the extra expense is warranted. On the other hand, many refiners wish only to make acceptable diesel, a less demanding application. For this, the less expensive molybdenum sulfides are adequate.

6.0 STRONG ACID CRACKING OF HYDROCARBONS

The cracking reaction results from attack of a strong acid on a paraffinic chain to form a carbonium ion (carbocation, e.g., R^+) (Figure 7-2). The reactions have been studied heavily since the 1920s; many of the most detailed studies are locked up in internal company reports and not available in the general literature, but there are several excellent treatises on this subject (e.g., Greensfelder et al., 1949).

Strong acids come in two fundamental types, Bronsted and Lewis acids. *Bronsted acids* are the familiar proton-containing acids; *Lewis acids* are a broader class including inorganic and organic species formed by positively charged centers. Both kinds have been identified on the surfaces of catalysts; sometimes both kinds of sites occur on the same catalyst. The mixture of Bronsted and Lewis acids sometimes depends on the level of water in the system.

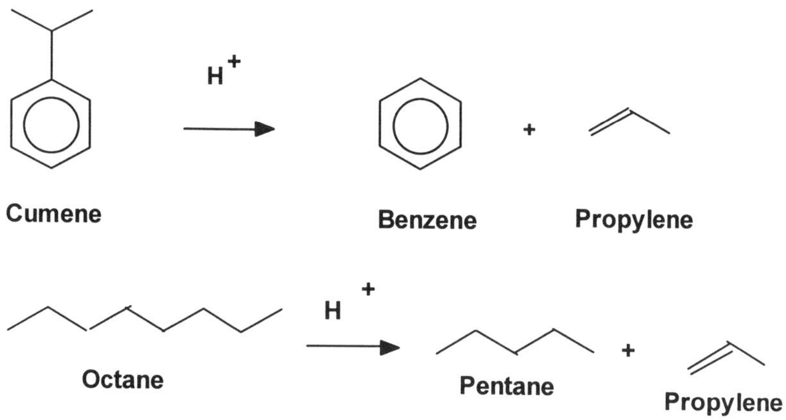

Figure 7-2. Strong acid cracking of model molecules.

6.1 Bronsted Acids

Examples of Bronsted acids are the familiar proton-containing species such as sulfuric acid (H_2SO_4). Acidity is provided by the very active hydrogen ion (H^+), which has a very high positive charge density. It seeks out centers of negative charge such as the pi electrons in aromatic centers. Such reactions are familiar to organic chemistry students, who are taught that bromination of aromatics takes place by attack of the bromonium ion (Br^+) on such a ring system. The proton in strong acid systems behaves in much the same way, adding to the pi electrons and then migrating to a site of high electron density on one of the carbon atoms.

6.2 Lewis Acids

The American chemist G. N. Lewis recognized that species other than proton donors behave as acids. These acids all have high positive charge densities. Examples are aluminum chloride ($AlCl_3$) and the bromonium ion (Br^+). Such strong positive species have become known as Lewis acids. This class obviously includes proton acids, but the latter are usually designated Bronsted acids in honor of the Danish chemist J. N. Brönsted (see above), who contributed greatly to the understanding of the thermodynamics of aqueous solutions.

In reactions with hydrocarbons, both Lewis and Bronsted acids can catalyze cracking reactions. For example, the proton in Bronsted acids can add to an olefinic double bond to form a carbocation. Similarly, a Lewis acid can abstract a hydride from the corresponding paraffin to generate the same intermediate (Figure 7-3). Although these reactions are written to show identical intermediates in the two reactions, in real catalytic systems the intermediates would be different. This is because the carbocations would probably be adsorbed on surface sites that would be different in the two kinds of catalysts.

7.0 SOLID, STRONG ACID CATALYSTS

Zeolites and amorphous silica aluminas provide the cracking function in hydrocracking catalysts. Both of these have similar chemistry at the molecular level, but the crystalline structure of the zeolites provides higher activities and controlled selectivities not found in the amorphous materials.

Chemists outside the catalyst field are often surprised that a solid can have strong acid properties. In fact, many solid materials have acid strength matching

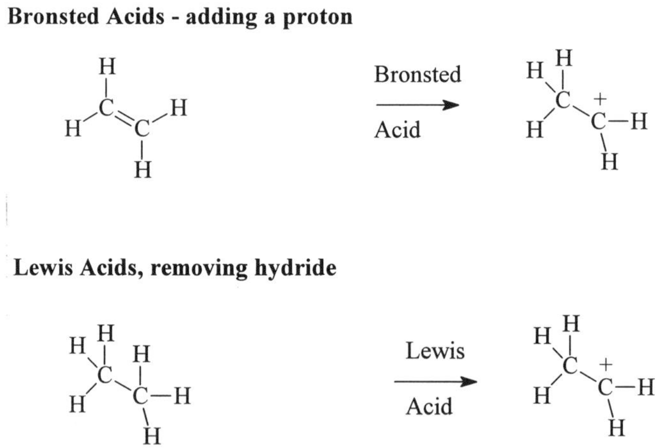

Figure 7-3. Cracking by Bronsted and Lewis acid catalysts.

that of concentrated sulfuric acid (Satterfield, 1991). Some specific examples are:

- Amorphous silica-alumina (SiO_2/Al_2O_3)
- Zeolites
- Activated (acid-leached) clays
- Aluminum chloride ($AlCl_3$) and many related metal chlorides
- Amorphous silica magnesia compounds (SiO_2/MgO)
- Chloride-promoted alumina ($Al_2O_3.Cl$)
- Phosphoric acid supported on silica gel (H_3PO_4/SiO_2)

Each of these is applied in one or more commercial catalysts in the petroleum refining industry. For commercial hydrocracking catalysts, only zeolites and amorphous silica alumina are used commercially.

7.1 Zeolites

Zeolites provide the cracking function in many hydrocracking catalysts, as they do in fluid catalytic cracking catalysts. The zeolites are crystalline aluminosilicates, and in almost all commercial catalysts today the zeolite used is faujasite. Pentasil zeolites, including silicalite and ZSM-5, are also used in some catalysts for their ability to crack long-chain paraffins selectively.

Typical levels are 25 to 50 wt. % zeolite in the catalysts, with the remainder being the hydrogenation component and a silica (SiO_2) or alumina (Al_2O_3) binder. Exact recipes are guarded as trade secrets.

Crystalline zeolite compounds provide a broad family of solid acid catalysts. The chemistry and structures of these solids are beyond the scope of this book. The field has been documented in many books, reviews, and articles (Rabo, 1976). What is important here is that the zeolites are not acidic as crystallized. They must be converted to acidic forms by ion exchange processes. In the process of doing this conversion, the chemistry of the crystalline structure is often changed. This complication provides tools for controlling the catalytic properties, and much work has been done on understanding and applying these reactions as a way to make catalysts with higher activities and more desirable selectivities (Avidan, 1993).

As an example, the zeolite faujasite crystallizes with the composition $SiO_2(NaAlO_2)_x(H_2O)_y$. The ratio of silicon to aluminum, expressed here by the subscript x, can be varied in the crystallization from 1 to greater than 10. What does not vary is the total number of silicon and aluminum atoms per unit cell, 192. For legal purposes, to define certain composition of matter patents, zeolites with a ratio of 1 to 1.5 are called type X; those with ratio greater than 1.5 are type Y.

Both silicon and aluminum in zeolites are found in tetrahedral oxide sites. The four oxides are shared with another silicon or aluminum (except that two aluminum ions are never found in adjacent, linked tetrahedral). Silicon with a plus four charge balances exactly half of the charge of the oxide ions it is linked to; since all of the oxygens are shared, silicon balances all of the charge around it and is electrically neutral. Aluminum, with three positive charges, leaves one charge unsatisfied. Sodium neutralizes this charge.

The sodium, as expected from its chemistry, is not linked to the oxides by covalent bonds as the silicon and aluminum are. The attraction is simply *ionic*, and sodium can be replaced by other cations by ion exchange processes. In extensive but rarely published experiments, virtually every metallic and organic cation has been exchanged into zeolites in studies by catalyst designers.

The most important ion exchanged for sodium is the proton. In the hydrogen ion form, faujasite zeolites are very strong acids, with strengths approaching that of oleum. Unfortunately, direct exchange using mineral acids such as hydrochloric acid is not practical. The acid tends to attack the silica alumina network, in the same way that strong acids attack clays in the activation processes developed by Houdry. The technique adopted to avoid this problem is indirect exchange, beginning with exchange of ammonium ion for the sodium. When heated to a few hundred degrees, the ammonium decomposes, forming gaseous ammonia and leaving behind a proton:

$$R^- NH_4^+ \rightarrow R^- H^+ + NH_3\uparrow$$

The step is accompanied by a variety of solid-state reactions that can change the zeolite structure in subtle but important ways. This chemistry and the related structural alterations have been described in many articles (Humphries *et al*, 1993). Because of its importance to the preparation of high-quality catalysts for hydrocracking and fluid catalytic cracking processes, the most important details are very closely guarded by catalyst manufacturers.

7.2 Amorphous Silica Aluminas

While zeolites provided a breakthrough that allowed catalytic hydrocracking to become commercially important, continued advances in the manufacture of amorphous silica aluminas made these materials competitive in certain kinds of applications. This was important, because patents controlled by Unocal and Exxon dominated the application of zeolites in this area.

Developments in amorphous catalysts by Chevron and UOP allowed them to compete actively.

Typical catalysts of this type contain 60 to 80 wt. % of the silica alumina, with the remainder being the hydrogenation component. The compositions of these catalysts are closely held secrets. Over the years, a broad range of silica/alumina molar ratios have been used in various cracking applications, but silica is almost always in excess for high acidity and stability. A typical level might be 25 wt. % alumina (Al_2O_3).

Amorphous silica-aluminas are made by a variety of precipitation techniques. The whole class of materials traces its beginnings to silica gel technology, in which sodium silicate is acidified to precipitate the hydrous silica-alumina sulfate; sulfuric acid is used as some or all of the acid for this precipitation, a mixed gel is formed (Iler, 1979). The properties of this gel, including acidity and porosity, can be varied by changing the recipe - concentrations, order of addition, pH, temperature, aging time, and the like. The gels are isolated by filtration and washed to remove sodium and other ions.

Careful control of the precipitation allows the pore size distributions of amorphous materials to controlled rather well, but the distributions are still much broader than those in the zeolites. This limits the activity and selectivity. One effect of the reduced activity has been that these materials have been applied only in making middle distillates: diesel and turbine fuels. At higher process severities, the poor selectivity results in production of unacceptable amounts of methane (CH_4) to butane (C_4H_{10}) hydrocarbons.

8.0 NITROGEN BASES AND HYDRODENITROGENATION

As would be expected, bases can bind with acidic sites and block access of target species. This is an important problem in hydrocracking, since petroleum contains nitrogen bases such as quinoline. As a general rule, hydrocracking processes provide a nitrogen removal function before the cracking catalyst bed. The exceptions to this rule are processes using amorphous silica alumina catalysts. Nitrogen poisoning is less of a problem with these, and hydrodenitrogenation can even be done with the hydrocracking catalyst.

8.1 Hydrotreating to Remove Nitrogen

Removal of nitrogen is carried out by hydrogenation. In fact, this is only one of a family of hydrotreating processes used in refineries.

Hydrodesulfurization (HDS) processes are used at several places in virtually every refinery, both to protect catalysts and to meet product specifications. On the other hand, hydrodenitrogenation (HDN) is commonly used only in conjunction with hydrocracking, to protect catalysts.

Other hydrotreating processes are used to saturate olefins and aromatics to meet product specifications or to remove metals from residual oils.

Virtually all of these processes rely on promoted MoS_2. Hydrodenitrogenation catalysts are usually nickel-promoted molybdenum sulfide (MoS_2), supported on alumina (Al_2O_3) (Topsoe *et al*, 1996). These nickel-moly catalysts are more active for hydrogenation than the corresponding cobalt catalysts. Good nickel moly hydrodenitrogenation catalysts are generally good hydrodesulfurization catalysts, as well.

In order to reduce the nitrogen in the feed to very low levels, hydrodenitrogenation catalysts are typically prepared with twice the level of metals found in hydrodesulfurization catalysts (Table 7-3). Phosphoric acid is sometimes added to hydrodenitrogenation compositions, both to stabilize the very concentrated impregnating solutions and to add acidity to the catalysts.

8.2 Hydrodenitrogenation Chemistry

Nitrogen removal by hydrodenitrogenation, particularly that of basic nitrogen compounds (Figure 7-4) (Chapter 3), requires complete saturation of the aromatic structure before nitrogen can be removed. This behavior is in sharp contrast to the removal of sulfur from thiophenic structures in hydrodesulfurization, in which the direct sulfur removal results in formation of

Table 7-3. Typical compositions of hydrodenitrogenation and hydrodesulfurization catalysts.

	HDN	HDS
Mo (as MoO_2)	25 wt. %	12 wt. %
Ni or Co, as NiO or CoO	5 wt. %	3 wt. %
H_3PO_4	1 wt. %	-
Al_2O_3	remainder	remainder

butadiene-like structures that are subsequently hydrogenated in rapid and easy reactions.

Detailed studies of the organic chemistry of hydrodenitrogenation (Cocchetto and Satterfield, 1981) show a complex network of reactions (Figure 7-5). Rate constants for all of the forward and reverse reactions were measured, with the highest rate path occurring through the network of intermediates (Figure 7-5). All of the intermediates were, however, observed in the reaction mixtures.

This chemistry, requiring saturation of the aromatic rings, has important implications for hydrodenitrogenation processes:

1. Hydrogen consumption will be high.
2. Heat release will be high because hydrogenation is exothermic.
3. The range of hydrodenitrogenation reactivities will be smaller than the range of hydrodesulfurization reactivities because the slow reactions are always ring saturations.
4. Aromatic saturation equilibria will inhibit hydrodenitrogenation rates as reactor temperatures approach 400°C (750°F). (Standard equilibrium constants for the hydrogenation of benzene are about 1 at about 290°C (550°F) and one atmosphere H_2, decreasing as the temperature rises; this equilibrium drives the formation aromatics in platinum reforming.)

Quinoline Pyridine Amine

Figure 7-4. Types of basic nitrogen compounds.

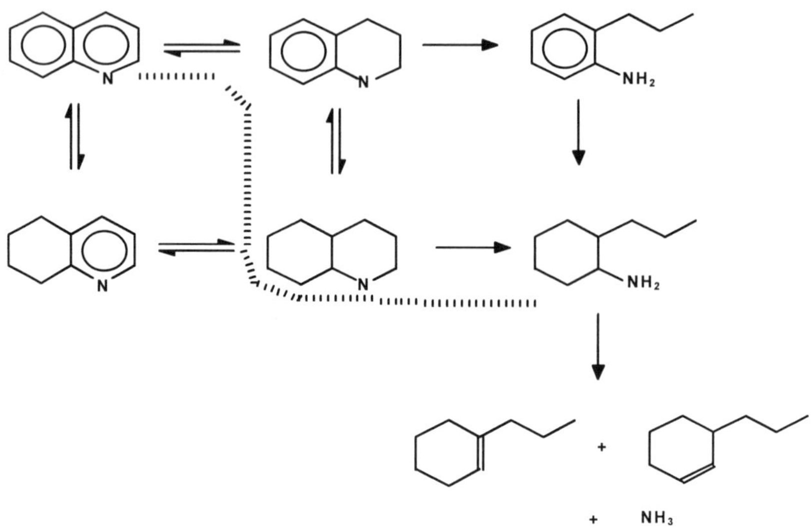

Figure 7-5. The quinoline hydrodenitrogenation network; the dotted line shows the route with the highest rate constants.

8.3 By-Product Ammonia

Hydrodenitrogenation generates ammonia when nitrogen is removed from aromatic structures in the oil. Although the ammonia is a base, it is much less effective in poisoning the acidic sites of zeolites. Ammonia binds much less tightly to the acidic sites than do quinoline bases.

This allows process designers to feed the gas and liquid products from hydrodenitrogenation directly to the hydrocracking reactors without separating the ammonia. Leaving the ammonia in the gas is an economic advantage, since ammonia removal would require cooling and reheating the stream, separating the gas to clean it, and so on.

9.0 POISONING BY COKE DEPOSITS

Hydrocarbons, especially aromatic hydrocarbons, can polymerize in the presence of strong acids to form coke. This coke is a complex polynuclear aromatic material that is low in hydrogen. Coke can deposit on the surface of a catalyst, blocking access to the active sites and reducing the activity of the catalyst. Coke poisoning is a major problem in fluid catalytic cracking catalysts, where coked catalysts are circulated to a fluidized bed combustor to be regenerated. In hydrocracking, coke deposition is virtually eliminated by the catalyst's hydrogenation function.

However, the product referred to as *coke* is not a single material. The first products deposited are tarry deposits that can, with time and temperature, continue to polymerize. Acid catalyzes these polymerizations. The stable product would be graphite, with very large aromatic sheets and no hydrogen. This product forms only with very high temperature aging, far more severe than that found in a hydrocracker (Marsh and Menendez, 1989). The graphitic material is both more thermodynamically stable and less kinetically reactive. This kinetic stability results from the lack of easily hydrogenated functional groups.

In a well-designed hydrocracking system, the hydrogenation function adds hydrogen to the tarry deposits. This reduces the concentration of coke precursors on the surface. There is, however, a slow accumulation of coke that reduces activity over 1–2 year period. Refiners respond to this slow reduction in activity by raising the average temperature of the catalyst bed to maintain conversions. Eventually, however, an upper limit to the allowable temperature is reached and the catalyst must be removed and regenerated.

Upper temperature limits are set by either metallurgy of the reactor, product quality considerations, or catalyst selectivity. Metallurgy considerations result from decreases in wall strength at higher temperatures. Product quality can be compromised at higher temperatures because thermal cracking reactions begin to compete with acid cracking as temperatures approach 400°C (750°F). Selectivity changes result from all the possible side reactions that become more important as the temperatures rise.

Coked catalysts can be regenerated by burning off the accumulated coke. This is done by service companies using rotary or similar kilns rather than leaving catalysts in the hydrocracking reactor, where the reactions could damage the metals in the walls. Removing the catalysts also allows inspection and repair of the complex and expensive reactor internals, discussed below. Regeneration of a large catalyst charge can take weeks or months, so refiners may own two catalyst loads, one in the reactor, one regenerated and ready for reload.

The thermal reactions also convert the metal sulfide hydrogenation functions to oxides and may result in agglomeration. Excellent progress has been made since the 1970s in regenerating hydrocracking catalysts; similar regeneration of hydrotreating catalysts is widely practiced.

After combustion to remove the carbonaceous deposits, the catalysts are treated to redisperse active metals. Vendor documents claim more than 95% recovery of activity and selectivity in these regenerations. Catalysts can undergo successive cycles of use and regeneration, providing long functional life with these expensive materials.

10.0 POLYNUCLEAR AROMATICS

Hydrocracking systems can be operated with recycle of the unconverted bottoms to increase overall conversions to desired products. In such processes, complex polynuclear aromatic (PNA) molecules such as coronene (Figure 7-6) accumulate in the recycled liquid. Even low levels of these molecules lend an intense red color to the liquid, leading to the nickname *red death*.

These polynuclear aromatic molecules are probably formed in the hydrocracking reactor, but they may have been in the feed and merely concentrated by the recycle. Whichever is the case, the concentrations of these molecules can exceed 1000 ppm, where they deposit on heat exchange surfaces and reduce heat transfer rates.

To avoid these problems, recycle hydrocrackers are designed with a bleed of 5–10% of the recycle material to keep the concentrations of the polynuclear aromatic molecules from exceeding a level of a few hundred parts per million.

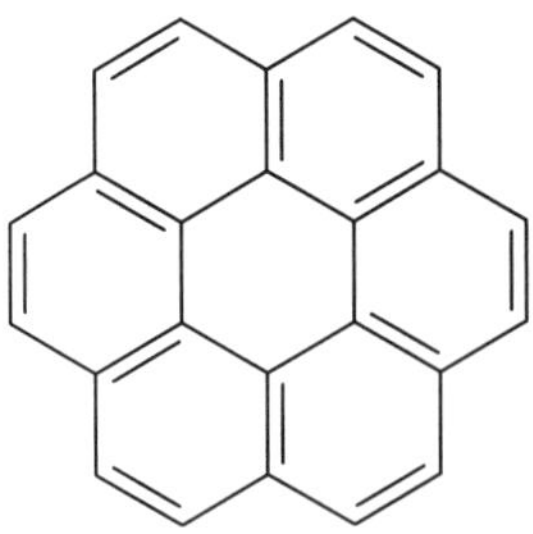

Figure 7-6. The polynuclear aromatic hydrocarbon coronene.

This bleed, being rich in hydrogen overall, is an excellent candidate for feed to a fluid catalytic cracker.

11.0 COMMERCIAL HYDROCRACKING

With the broad range of feedstocks available and the many process development groups that have explored hydrocracking since the 1940s, there is no one typical hydrocracking catalyst or process (Table 7-4). The majority of installed units are aimed at making midbarrel products (diesel and jet fuels). Within these, a great variety of catalysts is used. Nonetheless, temperatures and total pressures vary only within a fairly well-defined range (Ward, 1993; Hunter et al. 1994; Lewis et al., 1990).

11.1 Typical Process Configurations

The most general configuration for a hydrocracking process is a simple once-through system, with a hydrodenitrogenation reactor followed by the hydrocracking reactor. The two reactors can operate at the same pressure (Figure 7-7). After separation of the gas and liquids, distillation provides a set of products. Hydrogen gas is recycled and reused. This straightforward approach is

Table 7-4. Range of hydrocracking processes.

	Diesel/jet	Diesel/jet	Gasoline	Gasoline (recycle)
Catalyst	CoMo/SiAl	Pd/zeolite	Pd/zeolite	Pd/zeolite
Temperature, °C	260–370	260–370	290–380	290–380
°F	500–700	500–700	550–720	550–720
H2 pressure, psig	1500	800–1500	2200	2400
H2 consumption (scf/bbl)	400–1200	400–1200	>2000	>2000
HDN pretreat	NiMo	Integral	NiMo	NiMo
HC stages	1	1	1	2

widely used in hydrocrackers designed for making midbarrel (diesel and jet fuel) stocks.

Once-through systems can have one or two hydrocracking reactors, with the possibility of using fresh hydrogen in the second stage (Figure 7-8). This is valuable because the fresh hydrogen does not contain hydrogen sulfide (H_2S) and ammonia (NH_3) from the pretreater section, and it is not diluted by light hydrocarbons.

For higher conversions, some hydrocrackers use a recycle design. One very effective approach is the classic Unocal design (Figure 7-9). The product stream from the first hydrocracking stage is sent to a gas/liquid separator. First- and second-stage liquid products are mixed and distilled. The overhead is sent to product blending and the bottoms are sent to the second-stage reactor. This is not unlike the recycle configuration used in delayed cokers. As discussed earlier, polynuclear aromatic molecules can accumulate in the recycle stream and deposit

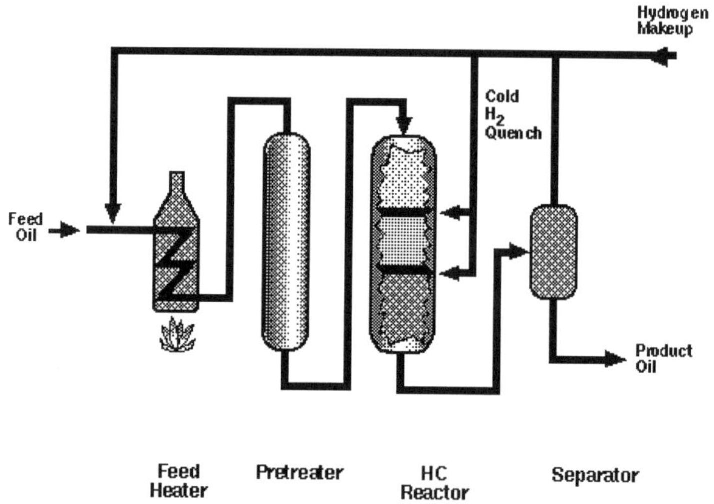

Figure 7-7. The once-through hydrocracking process.

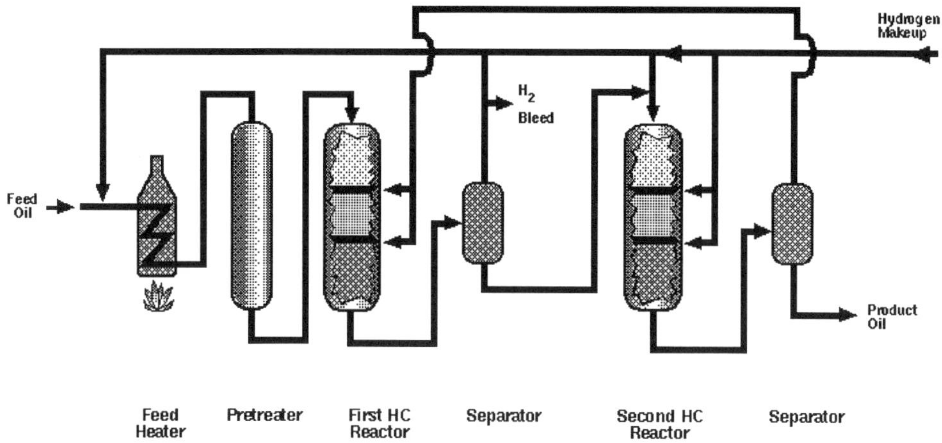

Figure 7-8. The two-stage once-through hydrocracking process.

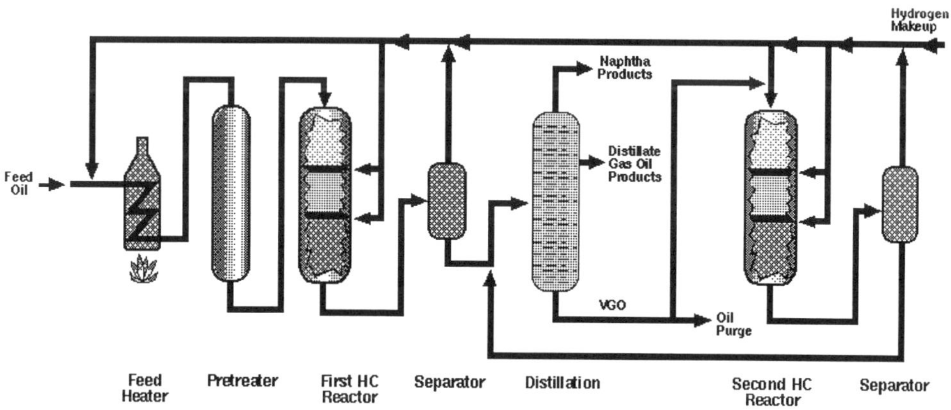

Figure 7-9. The two-stage once-recycle hydrocracking process.

on heater surfaces. This problem is solved by taking a small bleed stream. With amorphous catalysts, hydrodenitrogenation may be done in the first part of the hydrocracker.

Many of these processes are mechanically complex and can be represented by the use of simple block diagrams (Figures 7-7, 7-8, and 7-9). Many small steps are left out for the sake of simplicity, yet each of the steps must be carefully designed. For example, separation of oil and gases, with pressure reduction, requires several steps (1) hot, high-pressure separation, then (2) successively lower temperatures and pressures. This minimizes formation of foams and mists as dissolved gases come out of solution.

11.2 Temperature Control

Hydrogen is added at intermediate points in the hydrocracking reactors (Figures 7-7, 7-8, and 7-9). This is important for control of reactor temperatures. The mechanical devices in the reactor, called *reactor internals*, that accomplish this step are very important to successful processes.

Addition of hydrogen to olefins and aromatics is strongly exothermic, whereas catalytic cracking of paraffins to olefins is mildly endothermic. The net result of these thermal effects is a potential large adiabatic temperature rise, 225°C to 275°C (400°F to 500°F) as the oil passes through a hydrocracking reactor. Such a temperature rise is far too large for safe operation of a process.

Early process developers adopted the strategy of injecting cold hydrogen at various points in the reactor to cool the hydrogen-oil mixture. This approach is generally used today. The major problem faced in implementing hydrogen injection is to ensure that the oil is uniformly cooled. This is done by dividing the reactors into a series of separate catalyst beds or stages and adding the cold hydrogen between the beds. As the oil and gas emerge from a bed, they are collected and separated. Then hydrogen is injected and mixed thoroughly with the oil and gas. Finally, the two phase mixtures are carefully distributed across the top of the catalyst in the next lower bed.

If redistribution is done poorly, some areas of the bed will see more oil than others. This can lead to three levels of problems:

1. *Poor selectivity*: Ratios of hydrogen, oil, and catalyst outside design ranges will change the yield structures. Some parts of the bed will be hotter than others. Some fractions of the oil will be overcracked, others poorly converted. The result is reduced yields of desired products, increased production of light hydrocarbons, and lower overall conversions.

2. *Rapid catalyst aging*: Higher than desirable hydrogenation can increase local reactor temperatures markedly. Catalysts can sinter, losing surface area and activity and shortening run length.
3. *Hot spots*: When local reactor temperatures are well above 400°C (750°F), thermal cracking can become important. Thermal cracking produces olefins, which add hydrogen, releasing heat. This increases the temperatures further, and thermal cracking rates go up. These hot spots can easily reach temperatures higher than the safe upper limits for the reactor walls, and results can be catastrophic.

The separations and mixing involved in hydrogen injection are accomplished with static separators, mixers, and distributors. Gravity and gas flow drive their operation. Several generations of devices have been used, and today's reactor internals are quite efficient in cooling and redistributing the oil. This is an area of continuing research by process developers.

13.0 REFERENCES

Avidan, A. A. 1993. In Fluid Catalytic Cracking: Science and Technology. J. S. Magee and M. M. Mitchell, Jr. (Editors). Amsterdam, Elsevier, 1993. Chapter 1.

Boduszynski, M. M. 1987. Energy & Fuels 1: 2.

Boduszynski, M. M., 1988. Energy & Fuels. 2: 597.

Cocchetto, J. F., and Satterfield, C. N. 1981. Ind. Eng. Chem., Proc. Res. Dev. 20: 49.

Greensfelder, B. S., Voge, H. H., and Good, G. M. 1949. Ind. Eng. Chem. 41: 2573.

Humphries, A., Harris, D. H., and O'Connor, P. 1993. In Fluid Catalytic Cracking: Science and Technology. J. S. Magee and M. M. Mitchell, Jr. (Editors). Amsterdam, Elsevier, 1993. Chapter 2.

Hunter, M. G., Pappal, D. A., and Pesek, C. L. 1994. Annual Meeting. National Petroleum Refiners Association. Paper AM-94-21.

Iler, R. K. 1979. The Chemistry of Silica. John Wiley & Sons, Inc., Chapter 3.

Jahnig, C. E., Martin, H. Z., and Campbell, D. L. 1984. Chemtech 14(2): 106.

Keller, W.D. 1985. Clays. In Kirk Othmer Concise Encyclopedia of Chemical Technology. M. Grayson (Editor) Wiley Interscience, New York. p. 283.

Lewis, N. D., Dahlberg, A. J., Lammel, D. G., and Habib, M. M. 1990. Annual Meeting. National Petroleum Refiners Association. Paper AM-90-22.

Marsh, H., and Menendez, R. 1989. In Chapter 2 in Introduction to Carbon Science. H. Marsh (Editor). Butterworths, London. Chapter 2.

McEvoy, J. 1996. Chemtech. 26(2): 6.

Rabo, J. A. (Editor). 1976. Zeolite Chemistry and Catalysis. Monograph No. 171. Amercian Chemical Society, Washington, DC.

Satterfield, C. N. 1991. Heterogeneous Catalysis in Industrial Practice. 2nd Edition. New York, McGraw Hill. Chapter 7.

Speight, J. G. 1991. The Chemistry and Technology of Petroleum. 2nd Edition. Marcel Dekker Inc., New York.

Topsoe, H., Clausen, B. S., and Massoth, F. E. 1996. Hydrotreating Catalysis. Springer-Verlag, Berlin.

Ward, J. W. 1993. Fuel Process. Technol. 35: 55.

Chapter 8

Instability and Incompatibility of Petroleum Products

George W. Mushrush
Department of Chemistry
George Mason University
Fairfax
Virginia 22030

and

James G. Speight
Western Research Institute
365 North 9th Street
Laramie
Wyoming 82072

1.0 INTRODUCTION

As already noted, a *petroleum refinery* is a group of integrated unit processes (Figure 8-1), the use and combination of which are dependent upon the character of the crude oil (Speight, 1991; McKetta, 1992; Gray, 1994). Nevertheless, a refinery produces a variety of products from these processes that often require stabilization though hydrotreating. In this sense, the term *stabilization* refers to production of a product that, to all intents and purposes, does not undergo any further reaction when exposed to the air.

However, because of the complexity of the products, complete stabilization is rarely achieved and, as a result, the instability and incompatibility of petroleum products during storage and during refining operations have become major issues over the past two decades. Differences in crude oil composition and in the nature of the products from refinery operations have all served to emphasize the issue of instability and incompatibility in petroleum products.

Instability is defined as the inability of a petroleum product to exist for periods of time without a change in the product. Such changes might be reflected

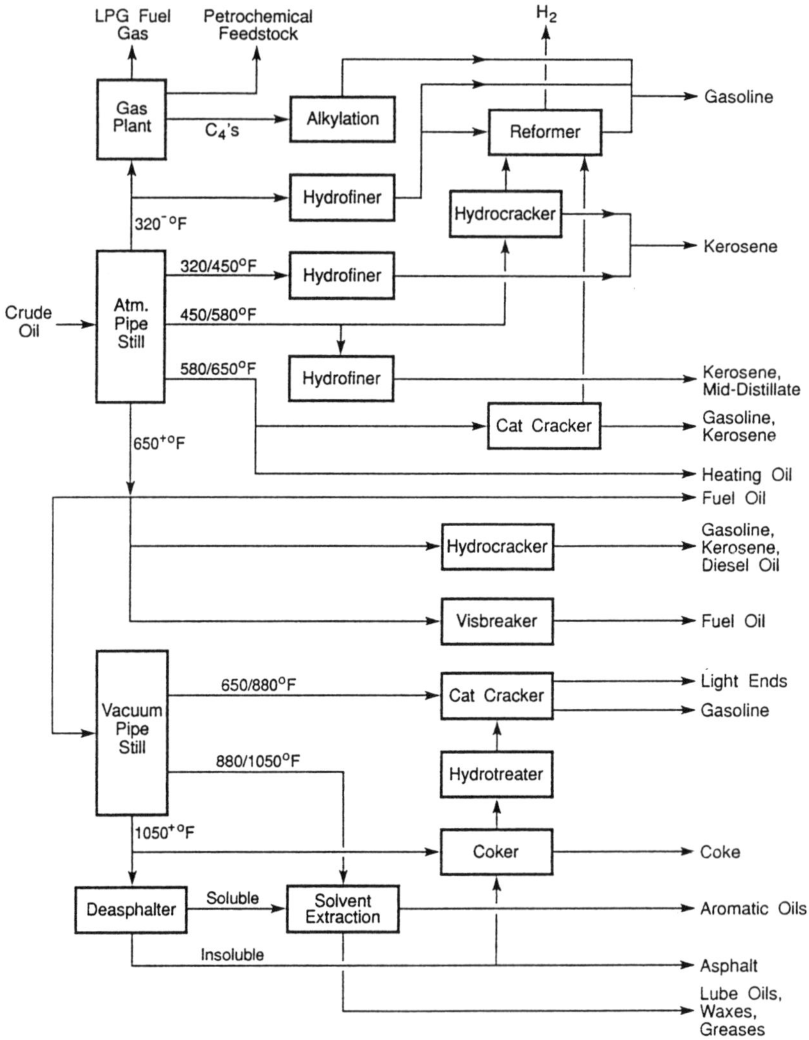

Figure 8-1. Schematic representation of a petroleum refinery.

as a noticeable change in color. The most common cause of instability is chemical in nature, such as the oxidation of petroleum and petroleum products resulting in the formation of oxidized products.

Incompatibility often refers to the *immiscibility* of petroleum products and of different crude oils and is often reflected in the formation of a separate phase after mixing. However, a product formed as a result of *instability* might also be *incompatible* with the original material, as might a product formed as a result of refining. Incompatibility also refers to the physical instability of the mixture.

The phenomenon of *instability* (thus, often variously referred to as *incompatibility* and more commonly described as *sludge formation*, *sediment formation*, or *deposit formation*) is a complex area of science and refers to the formation of a precipitate (or sediment) or phase separation during recovery operations, when two refinery streams or products are mixed, during product storage, and during refining operations (Wallace, 1964; Mushrush and Speight, 1995).

The term *instability* is often used in reference to the formation of color, sediment, or *gum* in the liquid over a period of time (Stavinoha and Henry, 1981; Hardy and Wechter, 1990; Ruzicka and Nordensen, 1990; Power and Mathys, 1992). *Storage stability* is a term used to describe the ability of the liquid to remain in storage over extended periods of time without appreciable deterioration as measured by gum formation and the deposition of insoluble material.

Thermal stability is defined as the ability of the liquid to withstand relatively high temperatures for short periods of time without forming carbonaceous deposits (coke) (Brinkman and White, 1981). There is also a tendency to define stability as the ability of the product to withstand long periods at temperatures up to 100°C (212°F) without degradation.

In general, the instability and incompatibility of petroleum products can be related to the heteroatom-containing (i.e., nitrogen-, oxygen-, and sulfur-containing) species in the feedstock (Figure 8-2) that, depending on the nature of the processing, appear in the products. In addition, recalling that all fractions of petroleum produce olefins as a result of thermal processes (Figure 8-3), the degree of unsaturation of the product (i.e., the level of olefins present) plays a role in determining instability or incompatibility.

Perhaps the true meaning of the term *incompatibility* is found when it is applied to specific refining processes (Section 4.0, this chapter). The application occurs when a product that is incompatible with (immiscible with or in soluble in) other products or the original oil is formed in the same reactor. An example is the formation of an insoluble solid phase that ultimately becomes coke during many thermal and catalytic operations. In the early stages of coke formation, the operative mechanism is considered to involve an initial *phase separation* of an

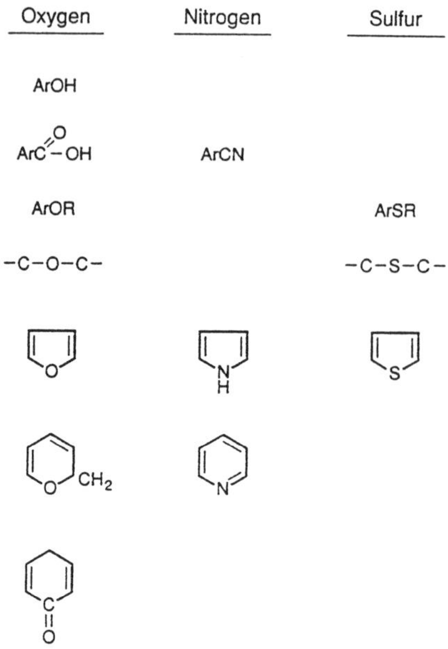

Figure 8-2. Nitrogen-, oxygen-, and sulfur-containing species in petroleum.

insoluble, solid, coke precursor prior to coke formation proper (Chapter 5) (Wiehe, 1993; Speight, 1994).

It has also been suggested (often, but not always, correctly) that the sediments and deposits originate from the inorganic constituents of petroleum. They may be formed from the inherent components of the crude oil (i.e., the metalloporphyrin constituents) or from contaminants ingested by the crude oil during the initial processing operations.

Sediments can also be formed from organic materials, but when reference is made to sediments and deposits, the inference is that these materials are usually formed from inorganic materials such as salts, sand, rust, and other contaminants that are insoluble in the crude oil and settle to the bottom of the storage vessel.

The *sediment* and other degradation products can have a marked influence on the properties of petroleum processes and products. Degradation products can affect color, volatility, flow characteristics of middle distillates, freezing point,

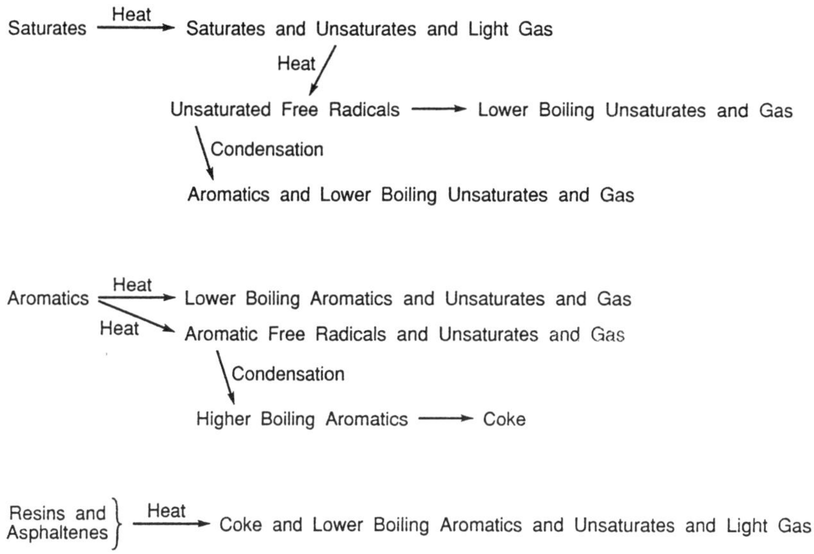

Figure 8-3. General representation of cracking reactions and products.

combustion properties, fuel oil properties, and rheological properties of asphalt (Traxler, 1961; Barth, 1962; Hoiberg, 1964; Broome and Wadelin, 1973).

Gum typically forms by way of a hydroperoxide intermediate that induces polymerization of olefins. The intermediates are usually soluble in the liquid medium. However, gums that have undergone extensive oxidation reactions tend to be higher in molecular weight and much less soluble.

The constituents of petroleum products can be defined as *incompatible* when sludge, semisolid, or solid particles are formed when such fuels are blended. The separate phases so formed can be either soluble or insoluble in the final blend. In the former case, they usually increase the viscosity of the product; in the latter case they settle out as semisolid or solid matter floating in the fuel or deposited on the walls and floors of containers.

Thus, the high molecular weight sediment that forms in liquid fuel is usually the direct result of autoxidation reactions. Active oxygen species involved include both molecular oxygen and hydroperoxides (Benson and Shaw, 1970). These reactions proceed by a free radical mechanism, and the solids produced tend to have increased incorporation of heteroatoms and are thus more polar and so increasingly less fuel soluble.

Fuel instability problems can be of two related types, high temperature short-term reactions, oxidative instability and ambient temperature long-term reactions, storage instability. Nitrogen-, oxygen-, and sulfur-containing compounds can be implicated in both types of instability. The synergistic effect of organic nitrogen and sulfur compounds with active oxygen species on the stability of middle distillate fuels is not well understood. Differences in fuel composition, reaction surface, quantity of dissolved oxygen, hydroperoxide concentration, and reaction temperature all contribute to the reported variations in fuel deterioration.

The soluble sludge/sediment precursors that form during processing or use may have a molecular mass on the order of several hundred daltons. For this soluble precursor to reach a molecular weight sufficient to precipitate (or to phase separate), one of two additional reactions must occur. Either the molecular weight must increase drastically as a result of condensation reactions leading to the higher molecular weight species, or the polarity of the precursor must increase (without any significant increase in the molecular weight) by incorporation of additional nitrogen, oxygen, or sulfur functional groups. In both cases, insoluble products will separate from the liquid medium.

In most cases a certain component in one of the fuels reacts with another component in the fuel with which it is blended. This chemical reaction results in the formation of a new product that, when soluble, affects the properties of the blend and, when insoluble, settles out in the form of semisolid or solid matter.

A special case of incompatibility of petroleum products involves *additives* incorporated in the various products.

Additives are chemical compounds intended to improve some specific properties of fuels or other petroleum products. Different additives, even when added for identical purposes, may be incompatible with each other, for example, react and form new compounds. Consequently, a blend of two or more fuels containing different additives may form a system in which the additives react with each other and so deprive the blend of their beneficial effect.

In general terms, studies of the composition of the incompatible materials often involve determination and definition of the *organic components* by *selective fractionation* analogous to that in the deasphalting procedure (Speight, 1991):

1. Heptane-soluble materials: often called *maltenes* or *petrolenes* (Chapter 1) (Speight, 1991);
2. Heptane-insoluble material, benzene- (or toluene-) soluble material: often referred to as *asphaltenes* (ASTM D-3279) (Chapters 1 and 4);

3. Benzene- and toluene-insoluble materials: referred to collectively as *carbenes* and *carboids* (Chapter 1)
4. Further definition of the benzene- or toluene-insoluble material as pyridine-soluble material (*carbenes*) and pyridine-insoluble material (*carboids*) (Chapter 1).

Carbon disulfide and tetrahydrofuran have been used in place of pyridine. The former (carbon disulfide), although having an obnoxious odor and being not much different from pyridine, is easier to remove because of its higher volatility. The latter (tetrahydrofuran) is not as well established in petroleum research as it is in coal-liquid related research. Thus, it is more than likely that the petroleum researcher will use carbon disulfide, pyridine or some suitable alternative solvent, although tetrahydrofuran has been known to form gum-like materials as a result of application to fractionation procedures.

Whichever solvent separation scheme is employed, there should be ample description of the procedure so that the work can be repeated not only in the same laboratory by a different researcher but also in different laboratories by various researchers. Thus, for any particular feedstock and solvent separation scheme, the work should repeatable and reproducible within the limits of experimental error.

The organic components in the sludge formed directly from petroleum are usually wax, wax-like substances, polar constituents, high molecular weight constituents (such as asphaltenes), and a variety of other organic (degradation) products (Table 8-1) (Por, 1992). In addition, there has been some success in determining the type of functional groups by the formation of functional group derivatives (Power and Mathys, 1992).

The degradation products formed during the long-term storage of crude oils have been identified (Mochida et al., 1986) and it has been postulated that the instability/incompatibility is a result of air oxidation, photooxidation, and condensation. Any high molecular weight polar substances that stay solid at ambient temperatures are so formed and consequently form sludge.

To summarize: part of the problem of incompatibility arises from the manner in which the fuels (petroleum and its product fractions, as well as other related liquids) react with oxygen during storage. The remainder of the problem may be due to the character of the individual constituents. Many of the constituents may be incompatible or immiscible with each other, resulting in phase separation, hence incompatibility.

Incompatibility in the middle distillate liquid fuel cut that is loosely defined as diesel and jet proves to be important to both commercial and military consumers as well as to the producer. Consumer uses, ranging from transportation to emergency auxiliary power supplies, require a stable, efficient fuel. This need

Table 8-1. Ranges of fractions usually found in sludge/sediment.

Fraction	Range wt. %
Heptane solubles	0–10
Heptane insolubles	10–60
Toluene insolubles	10–25
Wax	0–10
Water	0–50
Inorganics (as ash)	0–40

poses both chemical and physical limitations on what refiners can produce economically. Middle distillate fuels that are made from the refining process based on straight-run distillation show very few incompatibility problems. However, at present and in the future, problems for refiners will increase as the quality of the available crude decreases worldwide. This decrease, coupled with the inevitable future use of fuel liquids from both coal and shale sources, will exacerbate the present problems for the producers. Currently, producers try to maximize the gasoline fraction; thus, middle distillate is increasingly becoming a blend of cracked distillates such as light cycle oils. Care must be used in selecting these blending stocks so as to minimize incompatibility problems.

For example, diesel is a liquid product distilling over the range 150–400°C (300 - 750°F). The carbon number ranges, on average, from about C_{13} to about C_{21}). The chemical composition of a typical diesel fuel and how it applies to the individual specifications, API gravity, distillation range, freezing point, and flash point are directly attributable to both carbon number and compound classes present in the finished fuel (Tables 8-2 and 8-3).

The color test (ASTM D-1500) is a color comparison test. Consumers buying in bulk generally prefer a lighter colored product. As a fuel degrades, it generally becomes darker in color. This color change may or may not be related to degradation. Middle distillate fuels have a large proportion of the organonitrogen present as pyridines. Under acidic conditions, pyridines form charge transfer complexes that are highly colored. However, pyridines have never been implicated in incompatibility processes. Problems with incompatibility reactions in middle distillate fuels are associated with the formation of filterable sediments and gums.

Table 8-2. General ASTM specifications for the various types of diesel fuels.

Specification	Military[1]	No. 1-D[2]	No. 2-D[3]	No. 1[4]	No. 2[5]
°API	40.0	34.4	40.1	35.0	30.0
Sulfur %	0.5	0.5	0.5	0.5	0.5
B. p. °C	357	288	338	288	338
Flash point °C	60	38	52	38	38
Pour point °C	-6	-18	-6	-18	6
Hydrogen (wt.%)	12.5		-	-	-
Cetane No.	43	40	40	-	-
Acid no.	0.3	0.3	0.3	-	-

1. MIL-F-16884J
2. High-speed, high-load engines
3. Low-speed, high-load engines
4. Special-purpose burners
5. General-purpose heating fuel oil

Table 8-3. ASTM methods for determining fuel properties.

Specification	ASTM Method
°API gravity	D-1298
Total sulfur %	D-129
Boiling point °C	D-86
Flash point °C	D-93
Hydrogen (wt. %)	D-3701
Cetane number	D-613, D-976
Acid number	D-974

It has been known for a number of years that heterocyclic compounds containing oxygen, nitrogen, and sulfur are involved in the chemical processes of fuel instability (Mushrush, 1992). A few specific chemical reactions that may be important in instability have been identified. A particularly deleterious role is ascribed to the nitrogen and sulfur compounds present in fuels, but there is considerable disagreement about the specific species that are detrimental (Pedley et al., 1987, 1988).

The oxidation and condensation reactions of heterocyclic compounds in a medium as complex as a middle distillate fuel (Table 8-4) are most likely not isolated events. Thus, interactive effects between the active species in fuels need to be examined in greater detail. In many, if not all, fuels it is likely that interactive effects may actually define the incompatibility behavior. Little is known about thechemical mechanisms of such interactive effects in incompatibility reactions. Possible explanations include acid catalysis of condensation and/or oxidation reactions as well as interactive involvement in radical chain processes.

Table 8-4. Functional groups listed in order of incompatibility reaction in diesel fuels.

R–C=C- +	R–N– +	R–S– +	R–O–O–
organo- hydrocarbons	organo- nitrogen	organo- sulfur	organo- oxygen
alkenes	indoles	sulfonic acids	hydroperoxides
indenes	carbazoles	thiols	dissolved oxygen
naphthalenes	pyrroles	disulfides	carboxylic acids
monocyclics	quinolines	sulfoxides	peroxides
polycyclics	pyridines	sulfides	aldehydes

2.0 DISTILLATE PRODUCTS: INFLUENCE OF THE HETEROATOM FUNCTIONS

2.1 Oxygen

Practically all functional groups involving oxygen are present in hydrocarbon fuels. For the most part, aldehyde (–CHO), ketone (>C=O), alcohol (–OH), ester (–CO.OR), and ether (R–O–R′) groups have been shown to be present in low concentration and, more important, to be innocuous. The most deleterious oxygen compounds are the carboxylic acid (–CO_2H) and hydroperoxide (–O–O–H) species.

Carboxylic acids have been found to be detrimental in both fuels and model fuels. Carboxylic acids are implicated in incompatibility processes because they catalyze condensation reactions and thus rapidly increase both polarity and molecular weight. However, the most deleterious oxygen functional group is hydroperoxide, which can form by reaction of organic compounds with dissolved oxygen (Hazlett, 1980). It is observed that with time the aldehydes, ketones, and carboxylic acids form through the reaction of other active oxygen species. These compounds are the secondary reaction products of hydroperoxides.

Hydroperoxides are formed by the reaction of organic compounds with dissolved oxygen, but first, a *cautionary note*: all peroxides should be considered unstable. Even with precautions, unexpected and unexplained explosions can occur at moderate temperatures. It is probable that free radical reaction pathways can be altered with rapid evolution of low molecular weight gases resulting in large pressure surges. Reactions that involve anhydrous peroxides with fuels or in the presence of metals or metal ions should be approached with extreme caution. It is prudent to minimize the use of anhydrous peroxides or hydroperoxides as reactants. Hydroperoxide species are the primary oxidants present in any hydrocarbon fuel.

An increasing peroxide concentration is usually a predictor of the instability reactions to follow. Natural antioxidants present in fuels, primarily organosulfur compounds, keep peroxide concentration in check for short periods of storage time but are eventually overwhelmed. The ability of these compounds to play the role of antioxidants is probably what accounts for the induction period that is noted in instability reactions. The stability of all fuels could be radically improved by removing oxygen or lowering its concentration to less than 3 ppm. However, this would engender transportation and storage problems that might be too great to be overcome economically. Economical methods for lowering the dissolved oxygen content in storage are a fruitful area for further research. A nitrogen sparging, while easy in practice, would represent a large economic and environmental control burden but would probably be the cheapest of the alternatives.

Alkyl hydroperoxides are liquids; the lower members are soluble in both polar and nonpolar solvents. Low molecular weight hydroperoxides are explosive, but their thermal stability increases with increasing carbon number. Thus, the more stable peroxide compounds have a lower oxygen content. Alkyl hydroperoxides are considerably stronger acids than the alcohols. In reactions, homolytic, heterolytic, and oxidative cleavage of the peroxide bond can occur. In addition, peroxide formation is noted in many hydroperoxide reactions. In a basic medium, hydroperoxides give carbonyl compounds. Under acidic conditions, a few hydroperoxides have been observed to undergo rearrangement in addition to the traditionally accepted reaction pathways.

The autoxidation of hydrocarbons below 200°C (390°F) leads to extensive hydroperoxide formation. Hydroperoxides themselves are the main reactants in higher temperature regimes for subsequent autoxidation reactions. The O–O bond is the weakest bond in an alkyl or aryl hydroperoxide; the bond energy of the RO–OH bond is about 175 kJ/mol with thermal homolysis giving alkoxy and hydroxy radicals:

$$RO{-}OH \rightarrow RO + HO$$

At temperatures of about 100°C (212°F), alkyl hydroperoxides (but not allylic, $CH_2{=}CH.CH_2{-}$ or aryl, $C_6H_5{-}$) are stable in hydrocarbon solvents and it is the solvent itself that undergoes oxidation. When used as solvents, tetralin and cumene, for example, are oxidized to tetralin hydroperoxide and cumene hydroperoxide.

The thermal stability of tertiary alkyl hydroperoxides does not depend to any significant extent on the alkyl group present; thus *t*-butyl hydroperoxide and cumene hydroperoxide have similar chemical and thermodynamic stabilities. The stability of hydroperoxides is dependent on both the chemical environment and their concentration in a system. Stability based on chemical environment is more apparent than stability based on concentration. At a low concentration, $<10^{-3}$ M, of alkyl hydroperoxide, the reaction is first order in hydroperoxide, but at higher concentrations, $>10^{-1}$ M, it is second order.

The peroxide hydrogen bond of an alkyl hydroperoxide is weak and the reaction with other radicals in the system is exothermic. In high concentration, hydroperoxides are susceptible to induced decomposition by abstraction of the hydroperoxide hydrogen. However, the alkyl hydrogen atoms in tertiary groups are practically inert, because they are primary hydrogen atoms that are not activated by the presence of any other groups.

Primary and secondary alkyl hydroperoxides react quite differently. Secondary, unlike tertiary, alkyl hydroperoxides do not undergo an induced decomposition because of the formation of nonradical products, i.e., alcohols, ketones, and oxygen. Induced decomposition of hydroperoxides is catalyzed by trace amounts of metal ions, which are always present in fuel liquids. Reactions of hydroperoxides with metal ions have been well studied, but the synergism in complex systems such as fuels has been avoided. The reactions leading to the formation of hydroperoxides under mild conditions in fuels have been extensively reported (CRC, 1988). Systematic investigations of aliphatic hydrocarbon oxidation fuel mechanisms in the 1930s to the 1950s led to the conclusion that oxidation under mild conditions is a chain process. If the propagation steps are greater than those of termination, alkyl radicals will react with dissolved oxygen in the system:

$$R{-}H + M \rightarrow R + HM \qquad (1)$$

$$R + O_2 \rightarrow ROO \qquad (2)$$

$$ROO + R{-}H \rightarrow ROOH + R \qquad (3)$$

$$ROOH \rightarrow RO + OH \qquad (4)$$

$$ROOH + R'{-}H \rightarrow RO + R' + H_2O \qquad (5)$$

$$2ROOH \rightarrow ROO + RO + H_2O \qquad (6)$$

$$ROO + R' \rightarrow ROOR' \qquad (7)$$

$$2ROO \rightarrow ROH + R{=}O + O_2 \qquad (8)$$

$$2ROO \rightarrow ROOR + O_2 \qquad (9)$$

$$2R^1 \rightarrow R{-}R \qquad (10)$$

The rate constant for reaction (2) has been reported to be $k = 10^7–10^9$ M^{-1} sec^{-1} (Denisov, 1974). Consequently, if the oxygen concentration is at least 10^{-4} M, the peroxy radical, ROO, is present in much higher concentration than the alkyl radical, R. The peroxy radical is the least reactive of all the radicals generated. Thus a significant concentration of the alkyl peroxy radical is available for further reaction. In an oxygen-saturated pure hydrocarbon, with no peroxide species present, the chain process will not begin until a trace of the hydroperoxide radical forms.

Reactions (1) and (2) are important in starting the oxidation process but do not contribute significantly to the hydroperoxide concentration. In an oxygen-starved system, oxygen <5 kPa, reaction (2) becomes rate controlling.

2.2 Sulfur Species

Various types of sulfur species are present in fossil fuels and the organic chemistry of sulfur compounds plays an important role in crude oil or fuel stability (Reid, 1960; Danehy, 1968; Kharasch, 1961; Oae, 1977; Hartough, 1964). The influence of intermediate oxidation products such as sulfoxides and thiolsulfonates, along with the final product, a sulfonic acid, is equally important. It is probable that these intermediates and the final oxidized product hold the key to both sediment formation and the instability of middle distillate fuels.

Sulfides and thiols under oxidizing conditions produce the thiyl radical (RS) and thiyl radicals readily form disulfides (RS–SR), which, under mild or forcing oxidizing conditions, ultimately produce a sulfonic acid (RSO_3H). Other acidic species that might be proposed include sulfinic acids (RSO_2H) and sulfenic acids (RSOH). Sulfenic acids are generally unstable, and sulfinic acids are usually prepared through the corresponding ester by moderately strong oxidants. However, neither of these species is commonly observed and thus they have little significance for fuel properties.

Sulfides were first considered as inhibitors of deposit formation in gasoline and antioxidants in lubricating oils (Denison, 1944). Deposits formed in jet fuel in the presence of oxygen have been found to contain a greater percentage of sulfur than that present in the fuel itself. A similar result was observed for deoxygenated fuel doped with sulfides (Taylor and Frankenfeld, 1978, 1986). However some other reported results for sulfides are difficult to interpret. It has been reported that in Jet A fuel doped with butyl sulfide the amount of deposit formed differs from the amount formed when the same fuel is doped with pentyl sulfide (Daniel and Heneman, 1983). Bond energies for these two sulfides are practically identical. Unless impurities are present in the reagents, these two sulfides should give identical results. Experience shows that most commercially obtained sulfides contain small amounts of thiols.

A wide range of chemical interactions can occur with sulfides, including abstraction, fragmentation, combination, dimerization, addition to unsaturated bonds, displacement, and rearrangement. From the perspective of fuels, abstraction, combination, and addition to unsaturated bonds are the most probable reaction pathways, with radical annihilation by dimerization being a lesser pathway.

This is the problem with comparing model studies of systems employing fuels. Model studies assume a limited slate of radicals and alternative pathways available for reaction, which can be both a strength and simultaneously a weakness of model systems. However, general reaction pathways for both alkyl and aryl sulfides with the *t*-butoxy radical are recognized (Migita et. al, 1973):

$$RSCH_2CH_3 + t\text{-}C_4H_9O \rightarrow (11a) + (11b) + (11c) \quad (11)$$

$$RSCHCH_3\ (I) + RSCH_2CH_2\ (II) \quad (11a)$$

$$-CH_2{=}CH_2-\ (III) \quad (11b)$$
$$RS\ \ (IV) \quad (11c)$$

$$(I) + (I) \rightarrow (RSCHCH3)2 \quad (11d)$$

$$(I) + (II) \rightarrow RSCH_2CH(CH_3)SR \quad (11e)$$

$$(III) + (IV) \rightarrow (RSHCH_2)_2 \quad (11f)$$

$$(I) + (IV) \rightarrow RSCH(CH_3)SR \quad (11g)$$

Table 8-5. Mole percent conversion of hexyl sulfide at 120°C with *t*-butyl hydroperoxide in benzene.

	Mole % yield with time in minutes				
Sulfide products	15	30	60	120	180
Hexyl sulfoxide	74.8	85.9	81.6	81.9	80.7
Hexyl sulfone	1.2	1.4	3.0	3.4	4.0
Hexyl disulfide	0.3	0.5	0.5	0.5	0.6
Hexanal	0.2	0.2	0.3	0.2	-
Hexane	0.1	0.1	0.3	0.4	0.3
Hexene	0.1	0.1	0.1	0.3	0.2
Hexyl thiolsulfinate	0.1	0.1	0.1	0.1	0.1

The sulfur compound product slate for a model study involving the reaction of *t*-butyl hydroperoxide with hexyl sulfide in deaerated benzene solvent showed that the major product observed was hexyl sulfoxide (Table 8-5).

Other products included hexyl sulfone and hexyl disulfide. Minor products included hexyl thiolsulfinate, hexanal, hexene, and hexane. The hexyl sulfoxide could be produced result from several mechanisms, although the most likely mechanism is the reaction of the *t*-butyl hydroperoxide itself with the hexyl sulfide followed by a rapid proton transfer and O—O bond rupture. Expansion of the sulfur valence shell is probable in the processes involved in this step. Another mechanism could involve the attack of an oxygen-centered radical, i.e., peroxy on sulfur followed by a beta scission. In the presence of the very reactive peracids or peresters, sulfoxides and small amounts of sulfones are formed.

Only in the case of powerful peracids, i.e., trifluoroperoxyacetic acid, do sulfones form directly. The alkyl sulfone product then forms by a similar mechanism. The nucleophile attacks the sulfoxide as a sulfur electrophile. On subsequent reaction, peracids and esters ultimately form sulfonic acids from both aryl and alkyl sulfides. An interesting case involving a substituent effect involves phenyl alkyl sulfides. Aryl sulfides react and degrade on the basis of the substituent effect; that is a sulfonyl group protects the phenyl moiety, while an alkyl group is oxidized to a carboxylic acid.

Once formed, sulfoxides are quite stable under mild oxidizing conditions (Table 8-5). The major hexyl sulfoxide oxidation product, hexyl sulfone, varied from an initial 1.2% at 15 minutes to 4.0% at 180 minutes of reaction. This slight increase compared with the yield of hexyl sulfoxide at 180 minutes illustrated the resistance of an alkyl sulfoxide to oxidation by a hydroperoxide. These partially oxidized sulfur species, however, are not as stable as the starting hexyl sulfide. The mechanism for oxidation of a sulfoxide to a sulfone is believed to be similar to that for the sulfoxide formation. The formation of alkyl sulfones is a facile reaction only in the presence of a strong oxidant or when the reaction is catalyzed by transition metal ions.

The thiyl radical, RS, is the most reactive of the sulfur-centered radicals that can form in an oxidizing medium when thiols and disulfides are present. This radical can be generated from most sulfur species, but thiols serve as a convenient source. This sulfur-centered radical behaves quite differently from the corresponding oxygen-centered radical, the alkoxy radical. Alkoxy radicals produce both a cleavage product, a ketone, and a hydrogen abstraction product, an alcohol.

The chemistry of alkoxy and thiyl radicals is, however, quite different (reactions 12 and 13).

$$RS + R–H \rightarrow RS–H + R \quad (12)$$

$$(R)_3RS \rightarrow (R)_2R{=}S + R \quad (13)$$

For the thiyl radical, neither the thiol, to any significant extent, nor the thione has been observed to form by these reaction pathways. For reaction 12, the reason is thermodynamic. The reverse reaction is in theory favored by several kilojoules per mole. Only in cases involving very stable radicals is reaction 14 observed. In the case of reaction 13, thiones have never been observed from the simple thiyl radical cleavage.

The mechanism of thiol oxidation is still debatable. In the presence of oxygen, alkyl thiols are unreactive, with *t*-alkyl thiols being practically inert. Heavy metal ions, however, catalyze the oxidation of thiols. This is an area that is ripe for study: the interaction of metal ion catalysis with the hydroperoxide-initiated oxidation of sulfur species. In the presence of oxygen, alkyl thiols react through an intermediate hydrogen peroxide (reaction 14).

$$2RSH + O_2 \rightarrow RSSR + H_2O_2 \quad (14)$$

Results from the *t*-butyl hydroperoxide oxidation of alkyl thiols support the thesis that the *t*-butoxy or peroxy radical, generated from *t*-butyl hydroperoxide, abstracts the thiol hydrogen [reaction (15)]. Rapid dimerization [reaction (16)], then follows, generating the major product, an alkyl disulfide.

$$RS–H + t\text{-}C_4H_9O \rightarrow RS + t\text{-}C_4H_9OH \quad (15)$$

$$2RS \rightarrow S–S–R \quad (16)$$

Historically, in the presence of oxygen, the thiyl radical adds to olefins to give hydroxy sulfoxides as the major product and the sulfide as the minor product.

$$RCHCH_2SR + O_2 \rightarrow RCHCH_2S(R){=}O–O \quad (17)$$

In an acid-wash refining step, thiols can be oxidized to yield eventually very reactive sulfur species such as disulfides and sulfur dioxide [reaction (18)]:

$$2RSH + H_2SO_4 \rightarrow RSSR + SO_2 + 2H_2O \quad (18)$$

Sulfur dioxide then reacts with hydroperoxides [reaction (19)] in the system to produce a very reactive perester of a sulfonic acid:

$$ROOH + SO_2 \rightarrow ROSO_2OH \tag{19}$$

Reactions involving simultaneous addition and oxidation are quite common in both model systems and fuels.

A reaction sequence involving thiols with active olefins, such as styrene and indene, in the presence of hydroperoxides or molecular oxygen demonstrates these two cooxidation processes.

Whether the disulfide is a result of the oxidation reactions of thiols or was originally present in the fuel, the ultimate oxidation product is a sulfonic acid. A thiolsulfonate is stable if the substituent groups are long alkyl chains but unstable if they are short chains or aryl groups. With short chains, disproportionation to disulfides and thiolsulfonates is observed. Thiolsulfonates are quite stable, in contrast to sulfinyl sulfones. Disulfones are stable under ordinary conditions but can be cleaved by nucleophiles.

Under rigorous oxidizing conditions, a sulfonic acid is the expected product, but the situation is less clear under mild conditions. The generation of acidic species has important implications for the general area of fuel instability. Sulfenic and sulfinic acids are not usually formed by mild oxidation with hydroperoxides; thiolsulfinates, thiolsulfonates, and disulfones are observed under these conditions. Thus it is important to look at the sulfonic acid class of compounds present in both the crude and the finished fuel to understand the mechanism of fuel instability processes.

Model fuel studies can give misleading results with systems that are propagated by the generation of free radicals. Sulfur-centered radicals, under mild reaction conditions, can be quite high in concentration in both real fuels and model systems. Studies with high radical concentrations are usually conducted with solvents that are not susceptible to radical attack, leading to an abnormally high yield of sulfur radicals undergoing self-reaction. On the other hand, sulfur-centered radicals generated in a fuel matrix would be expected to be highly reactive, with no appreciable steady-state concentration under ambient conditions. There could be several explanations for this.

The very rapid disappearance of sulfur-centered radicals can be attributed to the dimerization reaction, which is thermodynamically favored under ambient conditions (H = -306 kJ/mol). This process is generally assumed to proceed at the diffusion-controlled rate constant. More probable would be combination reactions with olefins in the system, hydrogen abstraction reactions, and radical rearrangement. The reactions of any sulfur compound can then produce a thiyl radical, which by self-annihilation leads first to a disulfide and eventually to a sulfonic acid. Thus it is important to examine each functional class of sulfur compounds for sulfonic acid potential.

Substituted and condensed *thiophenes* are usually the most abundant heteroaromatic sulfur species in refined fossil fuels. There are several rapid and simple methods for the determination of both condensed thiophenes and other types of aromatic sulfur compounds based on ligand exchange chromatography coupled with gas chromatography–mass spectrometry. The thermal and reactive stability of the thiophenes approaches that of aromatic hydrocarbons, because of their resonance stability energies. The similarity to aromatic hydrocarbons also makes the direct gas chromatographic–mass spectrometric identification of these sulfur species difficult in a fuel matrix.

Thiophene compounds reported in the literature with reference to fuel instability have included thiophene, tetrahydrothiophene, alkyl-substituted thiophenes, and condensed thiophenes such as benzo- and dibenzothiophenes (Mushrush, et al., 1991). Oxidation reactions with peroxide species and the thiophenes, alkyl-substituted thiophenes, benzothiophene, and dibenzothiophenes have been reported. A thiophene ring system can survive moderate oxidizing conditions. However, if the ring is attacked, it generally breaks down to oxalic or other substituted carboxylic acids with the sulfur being converted to sulfuric acid. There is little information concerning the mild oxidation of thiophenes, especially in a radical environment. In the presence of low concentrations of hydroperoxides, no evidence of ring opening was observed (Mushrush et al., 1991).

Substituted thiophenes, such as benzothiophene, are the least reactive of the thiophenes under mild oxidizing conditions that would be encountered under ambient storage conditions of diesel or jet fuels. Under oxidizing conditions, when benzothiophene reacts with hydroperoxides, at the end of a 3-hour reaction more than 97% of the benzothiophene remained unreacted. In the presence of a hydroperoxide, benzothiophene yields minor amounts of the sulfone product and a trace of the sulfoxide product.

2.3 Nitrogen Species

Polar nitrogen compounds have two deleterious affects on both crude oil and fuels made from these crude oils. They poison oil refining catalysts and affect the quality of finished fuels (Chapters 3 and 7), since they can be implicated in product instability. Present knowledge has suggested that for some fuels, nitrogen heterocycles may play a causative role in the formation of insoluble sediments and gums under conditions of ambient and accelerated storage.

The nitrogen compounds that occur in finished fuels can be conveniently divided into a nonbasic and a basic fraction. The nonbasic nitrogen compounds include carbazoles, indoles, pyrroles, and quinolines. The basic fraction would include the pyridines and amines. In particular, nitrogen-containing aromatics, pyrroles, indoles, and carbazoles appear to be very harmful.

Nitrogen heterocyclic compounds that commonly occur in fuels include alkylated pyrroles in the gasoline fraction, along with alkylated pyridines, quinolines, tetrahydroquinolines, indoles, pryrroles, and carbazoles in middle distillate fuels (Table 8-6). The shale-derived fuel was subjected to severe hydrotreatment, and the petroleum fuel was a composite of 70% straight-run and 30% hydrocracked streams. Both of these fuels were marginally stable as defined by the formation of solid products.

Fuel instability is reflected by the concentrations of the polar nitrogen species present, and the carbazoles, indoles, and pyrroles are much more deleterious than other classes of polar nitrogen species. Practically all reports on the nitrogen class distribution in hydrocarbon liquids show that shale- and coal-derived fuels are higher in indoles, pyridines, and carbazoles, whereas petroleum fuels tend to be lower in these functional groups. However, the gasoline distillation cut appears to have a higher concentration of pyrrole-type nitrogen structures.

Currently, it appears that relatively nonbasic nitrogen heterocyclics, particularly those that contain alkyl groups in certain isomeric positions, may be the most troublesome compounds (Table 8-7). Thus, stereochemistry is quite important in determining the course of reactions in fuels.

Since both carboxylic and sulfonic acids are implicated in condensation reactions that lead to instability, the structure of the nitrogen compound would be expected to play a major role. Increasingly, we are led to believe that an interactive process between the basic and nonbasic nitrogen moieties is of major significance.

Pyridine compounds are thermally quite stable. Because they are easily protonated, they might seem to be likely candidates to be implicated in fuel instability reactions. However, pyridine compounds have not been observed in mass spectral analysis of any of the sediments. When pyridine compounds are used in model systems (Table 8-7), they do not induce further instability reactions. Most pyridine compounds either fit into group 4 ($<$3 mg solids) or do not induce formation of any sediment. Thus, they do not serve as either a reactant or a catalyst for further instability processes. The compounds that are found in fuels usually have alkyl substitutents in multiring positions. The carbon number of the alkyl chain is dependent on the fuel distillation range. For the shale fuel (Table 8-6), the nitrogen compound present in highest concentration was trimethyl-tridecylpyridine. The average chain length was found to vary from octyl (C_8H_{17}) to tetradecyl ($C_{14}H_{29}$). The alkyl chain length in the crude itself is much longer, but the alkyl side chain undergoes free radical rupture during thermal refining processes. It would not be wrong to consider the crude to consist of long alkyl chains containing random heteroatomic structures. Pyridine compounds are easily oxidized by strong oxidants to the corresponding pyridine *N*-oxide. Oxygen and hydroperoxides are not capable of oxidizing pyridine under ambient storage conditions, but at the considerably higher temperatures of nozzles or injectors both of these species could oxidize pyridines.

Substitution position on the pyridine ring markedly affects reactivity. Alkyl or alkenyl isomers in the 2- or 4-position result in compounds that are much more acidic than for the 3-isomer. These 2- and 4-isomers undergo the usual condensation reactions. However, in the 3-position, the resultant pyridines are quite stable.

Pyridines containing an active olefin group, i.e., 2-vinylpyridine, could be implicated in polymerization reactions. These compounds have many chemical similarities to styrene and other benzene homologs. The pyridine homologs are more reactive because of the labile N–H bond. Compounds such as 2-vinylpyridine would undergo reaction during the thermal refining and hydrotreating steps and not appear in the resultant fuels in any significant concentration. Alkyl-substituted pyridine compounds that have been subjected to hydrotreating yield alkyl-substituted piperdines. Piperdines show little tendency to react with weak oxidants.

Table 8-6. Nitrogen compound analysis up to 550 daltons for typical shale- and petroleum-derived jet fuels.

		% Nitrogen compounds	
Nitrogen compound class	Isomers	Shale	Petroleum
Carbazoles	98	1.3	0.5
Indoles	87	3.7	9.1
Pyridines/anilines	388	73.5	64.3
Pyrroles	42	0.6	1.1
Quinolines	66	4.1	10.4
Tetrahydroquinolines	248	9.9	3.2

Quinolines behave chemically in a similar fashion to pyridines. Model systems (Table 8-7) show that the quinolines fit in the least reactive category for sediment formation, also a pyridine similarity. On hydrotreating, quinolines can be easily converted to dihydroquninolines, tetrahydroquinolines, and decahydroquinolines. Dihydroquinolines undergo facile oxidation by air back to quinoline. In slightly acidic solutions, and with no oxygen present, dihydroquinolines will disproportionate to yield tetrahydroquinoline and quinoline. The quinoline-to-tetrahydroquinoline ratio is a measure of the severity of the hydrotreatment. Mass spectral analysis of sediments formed during storage or in thermal instability regimes do not show any significant incorporation of the quinoline nucleus.

Table 8-7. Relationship of sediment formation to organonitrogen compound structure.

Group 1 (>100 mg/100 ml)	*Group 2 (10–100 mg /100 ml)*
dodecahydrocarbazole	3-methylindole
2,5-dimethylpyrrole	2,5-dipropylpyrrole
2,4-dimethylpyrrole	pentamethylpyrrole
1,2,5-trimethyl pyrrole	2,3-dimethylindoline
Group 3 (3–10m g/100 ml)	*Group 4 (<3 mg/100 ml)*
2-methylpyrrole	2,6-dimethylpyridine
3-methylisoquinoline	3,5-dimethylpyrazole
1-methylpyrrolidine	2-methylquinoline
2,6-dimethylquinoline	2-methylpyrazine
2-benzylpyridine	2-methylpyridine
2-methylpiperidine	2-methylindole
pyrrolidine	3-methylpyridine
2,6-dimethylpyridine	4-methylpyridine
4-isobutylquinoline	4-methylquinoline
pyrrole-2-carboxaldehyde	1,2,3,4-tetrahydroquinoline

Quinoline is resistant to most mild oxidizing agents. Oxygen and peroxide concentrations commonly found in fuels apparently have little or no effect on the quinoline ring system. The use of stronger oxidants results in the formation of *N*-oxides. They follow the chemistry of the pyridine *N*-oxides with one major exception: the presence of a benzene ring allows electrophilic substitution. Electrophilic monosubstitution occurs at the positions adjacent to the ring junction. In this case a strong similarity to naphthalene chemistry is noted. Thus, temperature and reaction conditions both alter and govern the position of substitution. However, the position of the alkyl substitution does not seem to have any effect on reactivity with respect to instability.

Pyrroles show some chemical similarities but with significantly more differences from the pyridines and the other nitrogen functional groups. The major similarity is that of thermal stability. Differences include ease of oxidation of pyrrole and the facts that pyridine is basic and acid stable while pyrrole is practically neutral and readily polymerized, and pyridine is resistant to electrophilic reagents while pyrrole readily reacts. The ease of polymerization is shown by the reaction of pyrrole dissolved in ether containing a few drops of acid. Immediate polymerization results.

However, pyrroles that contain electron-withdrawing groups are either nonreactive or form simple salts with mineral acids. The oligmer product was referred to as *pyrrole black* in the early literature. The pyrroles that dimerize are characterized by those that have a free adjacent or ß position. This implies then that the reaction of 2,5-dimethylpyrrole is the result of autoxidation processes and not a simple polymerization reaction. The pyrrole oligmer, when heated, decomposes to pyrrole, indole and ammonia. The ease of oxidation for pyrrole compounds is illustrated by the rapid darkening of freshly distilled pyrrole. Thus the pyrroles that would be found in a finished fuel would be alkyl or aromatic substituted. This further suggest that feeds rich in pyrroles should probably not be subjected to an acid wash to remove basic nitrogen compounds, such as pyridines and amines. These basic moieties have been shown to be practically inert in finished fuels. Most fuels show a relatively low concentration of pyrrole structures, with most of the ones present being mono- or dialkyl-substituted short chains, methyl through pentyl. Short-chain multisubstituted pyrroles readily form peroxides upon exposure to air. Subsequently, these unstable peroxides would serve as a radical source for other oxidation reactions.

2,5-Dimethylpyrrole (DMP) has been the subject of many model system instability studies. 2,5-Dimethylpyrrole was probably employed because the reaction itself has been well defined. However, it bears little relationship to the actual processes occurring in fuels. Sediments generated from 2,5-dimethylpyrrole are not sensitive to the type of fuel, reaction temperature, solvent, or concentration. In short, these are good studies of the chemistry of 2,5-dimethylpyrrole and nothing more. When employed in instability studies, 2,5-dimethylpyrrole generated more sediment than any other nitrogen functional group.

Pyrroles readily react with aldehydes and ketones by 1,4 addition to give high molecular weight polymers. Aldehydes and ketones are the secondary reaction products from hydroperoxides. Thus their concentration can build to relatively high levels during ambient storage. But pyrroles are present in low concentration, so this reaction does not explain the preponderance of sediment formed.

Indole and *carbazole* are five-membered ring nitrogen compounds and, since carbazole can be considered a substituted indole, the two compound classes will be considered together (Table 8-7). To illustrate the sensitivity of alkyl substitution on reactivity, 2-methylindole was in the least reactive grouping. Indoles and pyrroles show similarities rather than differences. The similarities are that they are both slightly basic, resistant to substitution, similar in reactivity (C-3 of indole and C-2 of pyrrole are of the same order of reactivity), easy targets for electrophilic reagents, and they both form dimers and trimers.

Indole forms dimer salts in ether solvent that is slightly acidic. With aqueous acids, a mixture of dimers and trimers forms. Increasing polymerization is then the result of autoxidation reactions. Both indoles and carbazoles will readily form hydroperoxides. Autoxidation reactions readily occur with both 2,3-dialkyl- and 3-alkyl-substituted indoles. The 2-methylindole isomer is very stable and the very slight degree of reactivity (less than 1%) is still initiated at the C-3 position. Dimers of 2-methylindole cannot be synthesized directly. They must be synthesized by a reduction of the autoxidation product and once formed, they readily dissociate into the monomers.

Saturated carbazoles readily form a hydroperoxide. For example, both tetrahydrocarbazole and dodecahydrocarbazole form a similar hydroperoxide. Both the carbazole and indole hydroperoxides have a transitory existence and subsequently react further by a free radical process. The same compounds can form *trans* annular peroxides on exposure to air. Some of the indole hydroperoxide reaction products can be implicated in color body formation. 3-Hydroxyindole is a reaction product, derived from either 3-indole hydroperoxide or the oxidation of 3-methylindole with oxygen, that shows interesting chemistry. In slightly acidic media in air, a free radical reaction occurs at room temperature that results eventually in formation of the dye indigo.

Alkyl-substituted indoles show a wide range of reactivity. In slightly acid solutions, dimers and trimers of certain alkyl derivatives are formed. 2-Alkyl-substituted indoles do not condense to form dimers and trimers. 3-Alkyl-substituted indoles readily form dimers and trimers. However, mixed indole dimers and oligomers do form (Tables 8-8 and 8-9).

2.4 Condensation Reactions

Since both carboxylic and sulfonic acids are implicated in condensation reactions that lead to instability, the structure of the nitrogen compound would be expected to play a major role. Increasingly, the assumption that that an interactive process between the basic and the nonbasic nitrogen moieties is of major significance in interactive processes is strengthened. Possible explanations include acid catalysis of condensation or oxidation reactions, as well as interactive involvement in radical chain processes. Interactive processes between different types of nitrogen moieties have also been studied to a limited extent and may prove to be of great importance in middle distillate fuels (Frankenfeld et al., 1983; Watkins et al., 1989; Batts and Fahoni, 1991). The interactive studies involving the nitrogen compound 3-methylindole (3-MI) in the presence of a sulfonic acid, *p*-ethylbenzenesulfonic acid (EBSA), have proved to be of great value in predicting the chemistry of the formation of the deposits (Figure 8-4) in comparison with the

Table 8-8. Structure and reactivity of indoles.

Compound	Dimer and oligomers	No products	Extent of reaction
Indole	X		97%
1-Methylindole	X		94%
2-Methylindole		X	
3-Methylindole	X		92%
7-Methylindole	X		96%
3-Ethylindole	X		83%
3-*n*-Propylindole	X		62%
3-*n*-Butylindole	X		5%
3-*iso*-Propylindole		X	
3-*tert*-Butylindole		X	

Table 8-9. Reactivity of 2-methylindole with other indoles under mildly acidic conditions.

Indole compound	Formation of mixed oligomers	No oligomers
Indole	X	
1-Methylindole	X	
2-Methylindole		X
3-Methylindole	X	
7-Methylindole	X	
3-Ethylindole	X	
3-*n*-Propylindole	X	
3-*n*-Butylindole	X	
3-*iso*-Propylindole		X
3-*tert*-Butylindole		X

composition of the actual deposits formed in middle distillate fuels (Table 8-10) (Sundberg, 1970; Mushrush, et al., 1995).

3.0 RESIDUAL (NONDISTILLATE) PRODUCTS

3.1 Production and Classification

Fuel oil, also called *heating oil,* is a distillate product that covers a wide range of properties, and the specifications for such products are not as demanding as they are for other products (Hoffman, 1992).

No. 1 fuel oil is very similar to kerosene and is used in burners where vaporization before burning is usually required and a clean flame is specified. *No. 2 fuel oil* is often called *domestic heating oil* and has properties similar to those of diesel fuel and heavy jet fuel; it is used in burners where complete vaporization is not required before burning. *No. 4 fuel oil* is a light industrial heating oil and is used where preheating is not required for handling or burning; there are two grades of No. 4 fuel oil, differing in safety (flash point) and flow (viscosity) properties. *No. 5 fuel oil* is a heavy industrial fuel oil that requires preheating

Table 8-10. Elemental analysis for the reaction of 3-methylindole and ethylbenzenesulfonic acid (0.126 M 3-methylindole and 0.125M ethylbenzenesulfonic acid).

	Wt. %
Carbon	69.5
Hydrogen	5.9
Nitrogen	6.3
Oxygen	10.8
Sulfur	7.2
Empirical formula (structure 3)	$C_{26}H_{26}N_2O_3S$

before burning. *No. 6 fuel oil* is also a heavy fuel oil and is more commonly known as *Bunker C oil* when it is used to fuel ocean-going vessels; preheating is always required for burning this oil.

Residual fuel oil is obtained by blending residual products from various refining processes with suitable diluents, usually middle distillates, to obtain the required fuel oil grades. The residual products are obtained from (1) atmospheric distillation, (2) vacuum distillation, (3) visbreaking, and (4) catalytic cracking (Gruse and Stevens, 1960; Por, 1992; Gray, 1994).

Residual fuel oil is usually classified according to viscosity. For example, a light fuel oil is characterized by a viscosity (ASTM D-445) of 62 centistokes at 50°C (122°F), a medium-grade fuel oil by a viscosity of 175 centistokes at 50°C (122°F), a heavy fuel oil by a viscosity of 325 centistokes at 50°C (122°F), and extra heavy fuel oil by a viscosity exceeding 325 centistokes at 50°C (122°F).

Asphalt is also a nondistillate product and is, essentially, the bottom-of-the-barrel material that is used, for example, in road construction and waterproofing.

3.2 Properties

Fuel oil used for domestic purposes or for small heating installations would be of lower viscosity and lower sulfur content (Kite and Stephens, 1973; Walmsley, 1973). In large-scale industrial boilers, heavier grade fuel oil would be

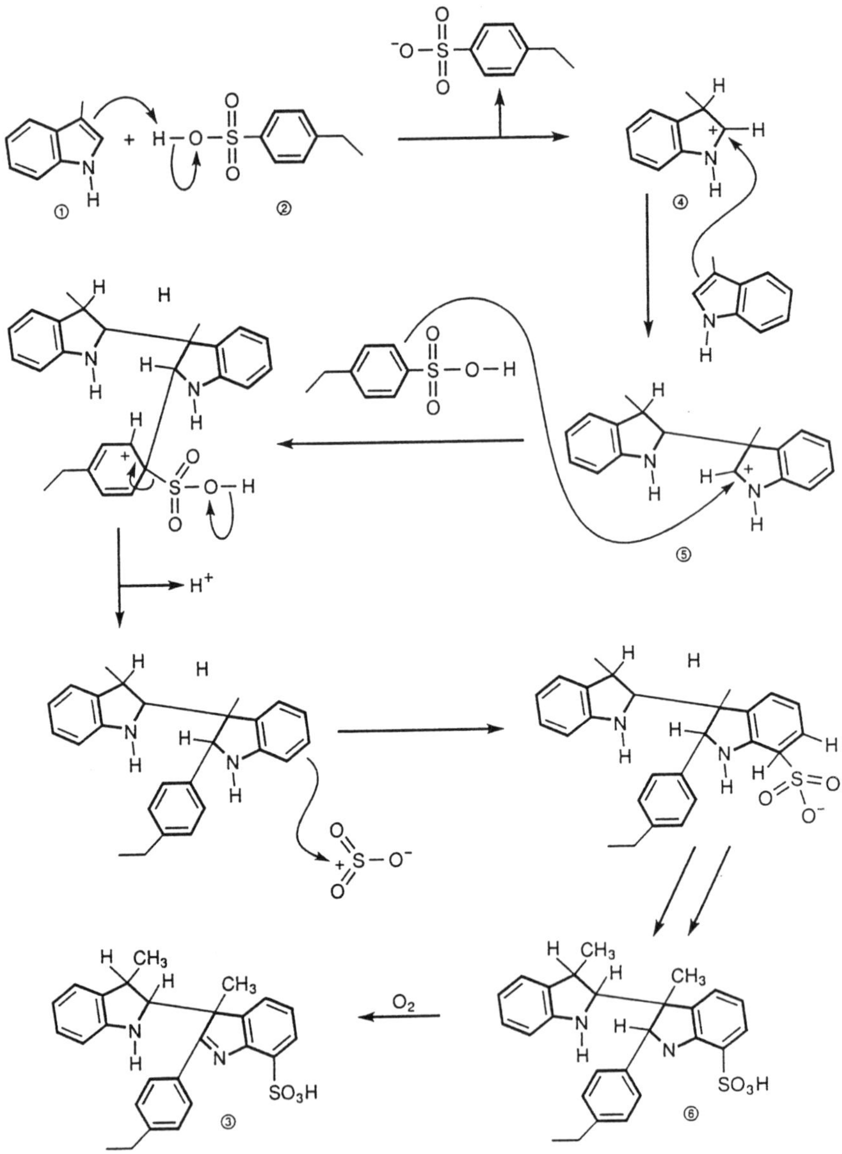

Figure 8-4. Predictive chemistry of the formation of deposits from liquid fuels.

used with sulfur content (ASTM D-1552 or ASTM D-129) requirements regulated according to the environmental situation of each installation and the local environmental regulations (Pope, 1973).

The *flash point* (ASTM-D 93) is usually limited to 60°C (140°F) minimum because of safety considerations. Asphaltene content (ASTM-D 3279), which is dependent upon the liquid hydrocarbon used for the precipitation (Figure 8-5) (Mitchell and Speight, 1973) (Chapter 1), carbon residue value (ASTM-D 189, ASTM-D 524), ash (ASTM-D 482), water content (ASTM-D 95), and metal content requirements are included in some specifications.

The *pour point* (ASTM-D 97), indicating the lowest temperature at which the fuel will retain its fluidity, is limited in the various specifications according to local requirements and fuel handling facilities: the upper limit is sometimes 10°C (50°F), in warm climates somewhat higher. Another important specification requirement is the heat of combustion (ASTM-D 240), the usually specified values being 10,000 cal/kg (gross) or 9400 cal/kg (net).

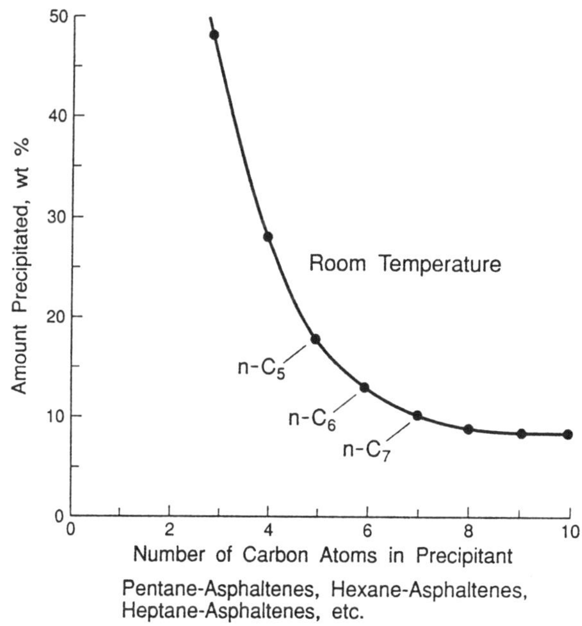

Figure 8-5. Variation of asphaltene yield with carbon number (molecular weight) of the hydrocarbon liquid.

Because of economic considerations, residual fuel oil has been replacing diesel fuel for marine purposes. Viscosity specifications had to be adjusted to the particular operational use and some additional quality requirements also had to be allowed for. The main problems encountered during the use of residual fuel oils for marine purposes are the stability properties (sludge formation) and even more so the compatibility properties of the fuel.

Blending of fuel oils to obtain lower viscosity values and mixing fuel oils of different chemical characteristics are sources of deposit and sludge formation in the vessel fuel systems. This incompatibility is observed mainly when high-asphaltene fuel oils are blended with diluents or other fuel oils of a paraffinic nature. Stability properties and compatibility of fuel oils, the mechanisms involved, suggested testing methods, and recommended remedies are described in detail later in this chapter.

3.3 Stability and Incompatibility

The specification requirements of the various fuel oil blends are attained by blending the residues obtained at the bottom of atmospheric distillation towers (straight run residues), vacuum residues, catalytic cracker residues, and thermal cracker/visbreaker residues with suitable diluents; the diluents are mostly middle or heavy distillates obtained in the various straight-run, vacuum, or cracking operations. The effect of the chemical composition of the diluents is of major importance in the stability and incompatibility/compatibility of the resultant fuel oil (Kite and Stephens, 1973; Por, 1992).

Residual fuel oil *stability* and *compatibility* (ASTM D-4740) are often affected by the severity of the cracking operation to which it had been subjected. Stability problems associated with residual fuel oils appear in the form of sludge and/or a solid deposit that settles at the bottom of a storage tank or plugs filters in fuel systems. Sludge and deposit formation is associated not only with product instability but also with incompatibility.

However, studies of the various mechanisms pertaining to the stability properties and the compatibility of residual fuel oils indicate that estimation of the peptization state of the asphaltene micelles would indicate well the degradation formation rate in residual fuels (Por et al., 1988).

The increasing need to produce more distillates from each barrel of crude oil necessitates employment of more severe conversion processes for increasing distillate yields. Alternative energy sources, diminishing the residual fuel demands, also contribute to the need to increase production of distillates at the expense of residues. More severe visbreaking conditions will have to be applied, adjusting the yield demand but resulting in residual fuels with less stability.

Vacuum residua, when serving as fuel oil base stocks rather than as visbreaker feedstock or asphalt feedstock, have to be blended with diluents because of viscosity requirements; compatibility problems may be expected in this case, especially when relatively high asphaltene content stacks are blended with diluents of an insufficient aromatic reserve.

Deposition of sludge or sediment may also be expected in blends of two or more residual fuels, forming a combined continuous phase of too low peptizing power. The use of fuel oil in marine engine systems, instead of marine diesel fuel or middle distillates, brought along problems of fuel system clogging by excessive formation of sludge and deposits, due in most of the cases to poor compatibility of the system components.

Stability and incompatibility/compatibility can be estimated in terms of (1) compatibility *spot tests* (ASTM D-4740), (2) *thermal stability* test data (ASTM D-1661), (3) existent and potential sludge formation (*hot filtration test*), (4) *asphaltene precipitation*, and (5) rate of *viscosity increase* of stored residual fuel oil samples at various temperatures with and without exposure to air.

In addition to the stability and compatibility test of residual fuel oil (ASTM D-4740), the other most frequently used stability test is the existent and accelerated dry sludge content determination of residual fuels (the *hot filtration test*). Hot filtration test results of up to 0.2% w/w are considered satisfactory and results above 0.4 wt. % indicate poor stability, but different values might be required, depending on the intended use of the product.

The tendency of a fuel oil to form degradation products may also be estimated by the *thermal stability* test (ASTM D-1661). There are certain limitations to this test, mainly with respect of viscosity: if the viscosity is not too high, this thermal stability test can be used; it indicates satisfactorily the fouling tendency of the tested fuel.

The tendency for free radical formation in various residues by different initiators of polymerization chain reactions has been investigated. The initiators used are heat, ultraviolet irradiation at wavelengths near the visible range (since carbon-hydrogen or saturated carbon-carbon bonds are not to be affected), accelerated oxidation, and catalytic action of metals. The free radical formation is then correlated with the behavior of the various types of residual fuels during storage and use.

The compatibility of various fuels with each other or with diluents can be determined (Por, 1992). Almost any diluent is compatible with catalytically cracked fuel oils or fuels of a high aromaticity; on the other hand, only diluents of an aromatic character are compatible with straight-run residues. This is of course to be expected, since diluents of a paraffinic character will precipitate asphaltene from fuel oil. The asphaltene content of catalytically cracked residues (a product of heavy vacuum gas oil) is low, and therefore it is compatible with almost any diluent.

Straight-run residues are assumed to be one of the more stable residual products; visbroken residues are the least stable. The thermal stability of catalytically cracked residues was found to be intermediate. Regarding blends with diluents, straight-run diluents are preferable in this respect to cracked diluents.

In order to prevent or to diminish the formation of degradation products in residual fuel oils and their blends, two controversial influences have to be considered. There is better compatibility in case of catalytically cracked residues and catalytically cracked diluents, the latter containing significant proportions of aromatic components. On the other hand, these products have poorer stability than straight-run residues. These straight-run residues and straight-run diluents have good stability, but there are problems in the compatibility of their blends.

These particular stability and compatibility properties can be explained by the availability (or nonavailability) of a sufficient aromatic reserve in conjunction with the micelle structure of the dispersed phase, which is a part of the colloidal solution characterizing the residual fuel oil structure.

Another area where incompatibility might play a detrimental role, during processing or in the product, is in *asphalt oxidation*.

The more polar species in a feedstock will oxidize first during air blowing (Speight and Moschopedis, 1979). After incorporation of oxygen to a limit, significant changes can occur in asphaltene structure, especially in terms of the incorporation of polar oxygen, which can influence molecular weight. Thus, the change in the character of the asphalt may be due not so much to oxidative degradation but to the incorporation of oxygen functions that interfere with the natural order of intramolecular structuring.

There is a potential for the incorporation of oxygen functions to enhance the ability of the asphalt to bond to the aggregate. But uncontrolled introduction of oxygen functions could result in the production of a poor grade asphalt where phase separation of the oxidized asphaltene may already have occurred. Or, should it occur in the product, the result can be pavement failure due to a weakening of the asphalt-aggregate interactions.

3.4 Nature of the Degradation Products

The instability or incompatibility of residual fuels is characterized by formation of separate phases, which usually takes one or more of the following forms: (1) precipitation of coke-like (carbonaceous) particles or inorganic impurities, both insoluble in aromatic solvents (ASTM D-4484); (2) separation of waxy materials; (3) precipitation of oxidation or polymerization products; (4)

precipitation of asphaltenes (ASTM D-3279); and (5) increase in the viscosity of the residual fuels during storage.

The asphaltene content of residual fuel oil is assumed to affect stability and compatibility (Por, 1992). Similarly, the chemical composition of residual fuel oil may well be assumed to be one of the main factors influencing the stability and the compatibility.

Asphaltenes are a complex mixture. The molecular structure of the constituents is not uniform (Chapter 1, 2, and 4) (Long, 1979, 1981; Speight, 1994) they and differ in their response and susceptibility to environmental influences as well as in their solubility in hydrocarbons. They exist in petroleum in the form of a micelle in which the nuclear components are surrounded by resin constituents, which, in turn, are surrounded by aromatic components. The dispersed and the dispersing phase are in equilibrium, which can be easily shifted in either direction (Chapter 4), and such a shift may result in a change of the residual fuel stability or compatibility.

The asphaltene contents of visbroken residua are usually much higher than those of the paraffinic vacuum residua, whereas their peptization values are significantly lower. Again, the asphaltene content increases with time, whereas the peptization values remain approximately constant. The higher asphaltene content of visbroken residua, in comparison with that of vacuum residua, is not surprising. It is a result of the concentration of the asphaltenes in the product. The asphaltenes themselves are not destroyed by the visbreaking process, but some of the other components are converted to distillates, which are separated from the nonvolatile residue.

The equilibrium of this system may be easily disturbed either by changes of temperature or by blending with less aromatic liquids. The degree of peptization of the asphaltene micelles becomes impaired and flocculation is initiated. This situation can be avoided by using blending stocks or diluents of sufficient peptizing power, that is, high aromaticity.

In the case of visbroken residua, the degree of peptization of the asphaltene micelles (which may be the inherent micelles as well as micelles formed during the cracking process) may also affect the stability of the fuel.

The susceptibility to oxidation processes is also associated with fuel instability, resulting in deposit formation (Por, 1992). During storage, oxidation and polymerization products may be formed that, if soluble in the fuel, will cause a viscosity increase. If these products are insoluble, they will settle out as highly viscous, semisolid, or solid phases.

Since the viscosity increase of residual fuel oil during storage indicates a change of the original properties, its rate of change may be considered as an estimate for the degree of their instability. In the case of residual fuel oil, the increase of viscosity as a result of degradation product formation is not only a quality problem but also an economic disadvantage, because the price of residual fuel oil is based on a reference viscosity. If the fuel oil viscosity rises, more distillate material is needed to reduce it to the reference viscosity level and this decreases the value of the more viscous fuel oil.

The rise in viscosity of stored fuel oil is generally due to degradation products that remain in solution; these products are formed during storage due to the temperature and/or the presence of aerial oxygen. If the chemical characteristics of the fuel oil are favorable to the dissolution of the degradation products, the viscosity of the fuel oil increases but sludge or deposit formation is often absent. Such fuel oil is often considered to be stable because there is no obvious sign of sludge or deposit. But, in fact, the degradation products are dissolved in the fuel oil.

An interpretation of the colloidal behavior of residual fuel oils, the peptization state of asphaltenes, or the tendency of the asphaltene micelles to remain peptized or to settle out and form sludgy deposits seems to be a suitable approach to the stability and compatibility problem related to residual fuel oil.

One approach is to base the instability/incompatibility of heavy residual fuels on the chemical composition as well as on the internal colloidal structure (Por, 1992) by defining a colloidal instability index (CII) as the ratio of the sum of asphaltenes and saturated oils to the sum of the peptizing resins and aromatic solvents:

Colloidal instability index = (asphaltenes + saturates)/(aromatics + resins)

The equilibrium of a well-peptized asphaltene system can easily be disturbed by (1) application of heat, (2) oxidation, (3) ultraviolet irradiation, and (4) addition of a paraffinic diluent. In each case, the chemical composition is altered and the aromaticity is decreased, causing a disruption of the equilibrium of the colloidal system.

As a result, asphaltene particles are deprived of their enveloping layers, which previously merged continuously with the consecutive layers. The micelle system becomes noncontinuous and the asphaltene cores are prone to agglomeration. Such a process leads to instability of residual fuel oils and formation of sludge and deposits at the bottom of storage tanks and of degradation products formation clogging lines, filters, and fuel systems.

Unstable fuels, originating mostly from cracking/visbreaking processes, tend to form free radicals, leading to polymerization chain reactions that result ultimately in formation of degradation products. When these are soluble in the fuel oil, or when the fuel oil, due to its chemical composition, is capable of dissolving them, its viscosity is increased.

When the degradation products do not dissolve in the fuel oil, either because of their nature or because the fuel oil is not able to dissolve them, they settle out as sludge or deposits. Often both processes occur simultaneously, with part of the degradation products being dissolved (with an accompanying increase in fuel oil viscosity) and part remaining insoluble and forming sludge and deposit.

Degradation products formed because of the incompatibility of some of the fuel oil components are associated with a different mechanism. When two or more fuels, incompatible with each other, are blended, the degradation products are formed by a chemical reaction between two or more chemical groups present in the fuel oil components. In most cases, the products either settle as a sludge or float in the fuel oil, forming a heterogeneous solution.

Incompatibility in residual fuel oil is often due to a high asphaltene content of one of the components in the blend, especially when such a component is blended with another fuel of a predominantly paraffinic nature or when the high-asphaltene component is blended with a paraffinic diluent. In this case the asphaltenes are precipitated by the paraffinic constituents (Speight, 1991).

However, not only the concentration of the asphaltenes but also their peptizability is of importance in this respect. Most estimates of the potential for compatibility and/or incompatibility in residual fuels are based on determinations of sludge and suspended solids formation rates on blending with diluents.

In the compatibility *spot test* (ASTM D-4740) a drop of the preheated fuel oil sample is placed in an oven at 100°C (212°F) and after 1 hour the spot is examined for evidence of precipitation; at this stage a stability estimate is obtained. As a second step, a blend consisting of equal amounts of the fuel oil and the blending stock is handled similarly: a compatibility estimate is so obtained.

In an alternative test (ASTM D-2781), compatibility ratings are based on numerical values for a reference spot obtained under the particular testing procedure in which a drop of the fuel oil and diluent blend is allowed to spread on a chromatographic paper of a specified grade. The exact procedures vary according to the particular method used. Assigned estimates from 1 to 5 indicate the homogeneity of the obtained spot, 1 being the most homogeneous spot, indicating the best compatibility, and 5 representing a very dark spot surrounded by light-colored areas, indicating very poor compatibility.

Another procedure, sometimes called the *xylene equivalent test*, is based on determining the proportion of xylene in a blend with iso-octane necessary for obtaining a homogeneous spot on a chromatographic paper when blended with an equal quantity of the tested fuel oil sample; the lower the proportion of xylene in the diluent, the better is the compatibility of the tested fuel.

A modified spot test for determining the compatibility of fuel oil is based on using a diluent composed of equal proportions of *n*-heptane and toluene or of two types of gas oils, one a straight-run (paraffinic) and the second a catalytically cracked (olefinic/aromatic) gas oil. The spots obtained by drops of the fuel oil and diluent blend on a chromatographic paper are evaluated as with the other spot tests, but this modified test requires a higher temperature (usually 63°C; 145°F) before asphaltene deposits are indicated.

The test for sludge formation (*hot filtration test*) is useful for stability as well as for compatibility estimations. Final fuel oil blends yielding values below 0.3% in this test would indicate good stability properties as well as acceptable compatibility.

Data on the peptization state of asphaltenes are useful in compatibility estimations. High peptizability, as indicated by the lower values obtained in the procedure, points to a stable colloidal system in which micelles are homogeneously integrated. Such a system would exhibit good compatibility.

Viscosity increases during fuel oil storage at various temperatures, without and with contact with air, are a good estimate of the compatibility of such blends. Viscosity increases indicate formation of degradation products that have been dissolved in the fuel.

Because not only incompatibility but also instability of fuels are responsible for such degradation product formation, care should be taken when interpreting such results. When samples are kept out of contact with air, the effect of instability is reduced. In this case incompatibility would be the main cause of such a viscosity increase of the fuel oil blend during storage.

4.0 INCOMPATIBILITY IN REFINING OPERATIONS

There are two general routes to petroleum refining: (1) removing carbon as coke, often referred to as carbon rejection, and producing lower boiling products as overhead (distillate material); and (2) adding hydrogen by hydroprocessing, by which the yield of coke is reduced in favor of lower boiling products (Figure 8-6) (Chapters 5, 6, and 7).

The carbon rejection mechanism is, by definition, a means of removing a low-hydrogen product and allowing distribution of the inherent hydrogen among the lower boiling products. In addition, carbon rejection is, by definition, virtually a process by which the incompatibility of one product with the others is the means by which the process succeeds.

Once a refinery feedstock has been defined sufficiently well to indicate the primary methods of refining, it is the nature of the processing operations that dictates the product slate. But it is the nature of the feedstock and the products that dictates the potential for instability and/or incompatibility.

Separation of solids during refining can occur in a variety of processes, either by intent (such as in the deasphalting process) or inadvertently when the separation is detrimental to the process. Thus, separation of solids occurs whenever the solvent characteristics of the liquid phase are no longer adequate to maintain polar and/or high molecular weight material in solution.

Examples of such occurrences are (1) *asphaltene separation*, which occurs when the paraffin content or character of the liquid medium increases; (2) *wax separation*, which occurs when there is a drop in temperature or the aromaticity of the liquid medium increases; (3) *sludge/sediment formation*, which occurs when the solvent characteristics of the liquid medium change so that asphaltic or wax materials separate; (4) *coke formation*, which occurs at high temperatures and commences when the solvent power of the liquid phase is not sufficient to maintain the coke precursors in solution; and (5) *sludge/sediment formation in fuel products*, which occurs because of the interplay of several chemical and physical factors.

Perhaps the most obvious occasion for incompatibility to occur during refining is in *thermal processes*.

When petroleum fractions are heated to temperatures over 350°C (660°F), the thermal or free radical reactions start to give cracking of the mixture at

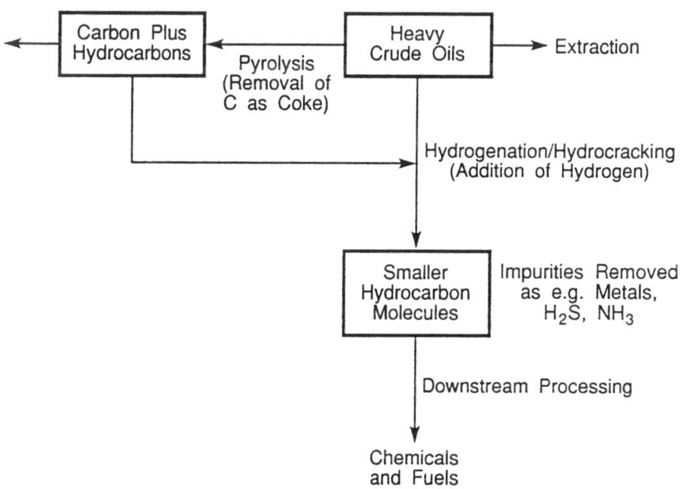

Figure 8-6. Schematic representation of heavy oil processing options.

significant rates (Chapter 5). Thermal conversion does not require the addition of a catalyst; therefore, this approach is the oldest technology available for residue conversion. The severity of thermal processing determines the conversion and the product characteristics.

The chemistry of refining processes is often represented by simple chemical equations but is, in reality, a complex sequence of chemical reactions (Chapter 5) (Speight, 1991; Gray, 1994). The complexity of the feedstock influences the complexity of the products. Thus, there are many unknowns in the chemistry and physics of incompatibility during refining operations.

The thermal treatment of residua ranges from mild treatment for reduction of viscosity to ultrapyrolysis for complete conversion to olefins and light ends (Speight, 1991; Gray, 1994). The higher the temperature, the shorter the time to achieve a given conversion. The severity of the process conditions is the combination of reaction time and temperature to achieve a given conversion.

If no side reactions occur, very long times at low temperature should be equivalent to very short times at high temperature. Thermal reactions, however, can give rise to a variety of different reactions so that selectivity for a given product changes with temperature and pressure. Mild- and high-severity processes are frequently used for processing of residue fractions. More extreme conditions are used commercially only for cracking ethane, propane, butane, and light distillate feeds to produce ethylene and higher olefins.

The main limitation on straight thermal processing is that the products can be unstable. Thermal cracking at low pressure gives olefins, particularly in the naphtha fraction. These olefins give a very unstable product, which tends to undergo polymerization reactions to form tars and gums. The heavy fraction can form solids or sediments, with feed composition also determining the maximum conversion allowable (Allan et al., 1983). However, before progressing to the thermal processes, there is one process in which incompatibility is used to beneficial effect in product separation, and that is the deasphalting process.

As examples of processes in which compatibility might occur, it is appropriate, for the present purposes, to select the visbreaking process (*a carbon rejection process*) and the hydrocracking process (*a hydrogen addition process*) as occur in an integrated refinery. The processes operate under different conditions and have different levels of conversion.

The effect of the higher temperatures on the separation of asphaltic material is largely unknown, although there are indications (Rao and Serrano, 1986; Schabron and Speight, 1997) that higher temperatures discourage aggregation. In the high-temperature reactor, it might be presumed that it is the precursors of coke that separate from the liquid phase and not the original or unchanged asphaltenes.

4.1 Visbreaking

The visbreaking process is a means of reducing the viscosity of heavy feedstocks by *controlled* thermal decomposition (Chapter 5) (Speight, 1991). However, the process is often plagued by sediment formation in the products. This sediment, or sludge, must be removed if the products are to meet fuel oil specifications.

The process uses mild thermal cracking as a relatively low-cost and low-severity approach to improving the viscosity characteristics of the residue without attempting significant conversion to distillates. Low residence times are required to avoid polymerization and coking reactions, although additives can help to suppress coke deposits on the tubes of the furnace (Allan et al., 1983; Gary and Handwerk, 1984).

The reduction in viscosity of the unconverted residue tends to reach a limiting value with conversion, although the total product viscosity can continue to decrease (Gray, 1994). Conversion of residue in visbreaking follows first-order reaction kinetics (Henderson and Weber, 1965). The minimum viscosity of the unconverted residue can lie outside the range of allowable conversion if sediment begins to form (Rhoe and de Blignieres, 1979). When transportation of visbreaker product by pipeline is the process objective, a diluent such as gas condensate can be added to achieve a further reduction in viscosity.

The high viscosity of the heavier feedstocks and residua is thought to be due to entanglement of the high molecular weight components of the oil and the formation of ordered structures in the liquid phase. Thermal cracking at low conversion can remove side chains from the asphaltenes and break bridging aliphatic linkages. Such events are sufficient to reduce the structural integrity of the liquid phase and lead to a reduction in viscosity, which is accompanied by a reduction in the pour point.

There are several chemical models (e.g., Schucker and Keweshan, 1980; Gray, 1994 and references cited therein) for the thermal decomposition of asphaltenes that can be described in composite form and are worthy of consideration. Using asphaltene models as a guide (Speight, 1991), the asphaltene nuclear fragments become progressively more polar as the paraffinic fragments are stripped from the ring systems by scission of the bonds (preferentially) between the carbon atoms alpha and beta to the aromatic rings.

The polynuclear aromatic systems, being denuded of the attendant paraffinic substituents, are somewhat less soluble (Bjorseth, 1983; Dias, 1987, 1988) in the surrounding hydrocarbon medium than the *unreacted parent systems*, not only because of the solubilizing effects of the alkyl substituents but also because of the enrichment of the liquid medium in paraffinic constituents.

Thus, the stability of visbroken products becomes an issue since these products now contain polar species that have been denuded of some of the alkyl chains and which, on the basis of solubility, might be more rightly called carbenes and carboids (Chapter 1), but where an induction period is required for phase separation or agglomeration to occur. Such products might, initially, be *soluble* in the liquid phase, but after the induction period (a pregnant pause), cooling, and/or diffusion of the products, incompatibility (phase separation, sludge formation, agglomeration) occurs.

4.2 Coking

Coking, as the term is used in the petroleum industry, is a process for converting nondistillable fractions (residua) of crude oil to lower boiling products and coke. Coking is often used in preference to catalytic cracking because of the presence of metals and nitrogen components that poison catalysts. There are actually several coking processes: *delayed coking*, *fluid coking*, and *flexicoking* as well as several other variations (Chapters 5 and 6).

A premise similar to that as outlined for the visbreaking process also applies in coking processes, in which the changing nature of the feedstock can cause incompatibility of the products and, thence, be an adverse forerunner to the conversion by a coking process.

4.3 Hydrotreating and Hydrocracking

In contrast to the visbreaking process, in which the general aim is to produce products for use as fuel oil, the hydrocracking process is employed to produce a slate of products for use as liquid fuels.

Nevertheless, the decomposition of asphaltenes is, again, an issue, and just as models consisting of large polynuclear aromatic systems are inadequate to explain the chemistry of visbreaking, they are also of little value for explaining the chemistry of hydrocracking.

The presence of hydrogen changes the nature of the products, especially the coke yield, by preventing the buildup of precursors that are incompatible in the liquid medium and form coke (Chapter 5) (Magaril and Aksenova, 1968; Magaril and Ramazaeva, 1969; Magaril et al., 1970). But precisely how asphaltenes react with the catalysts is open to much more speculation.

Incompatibility is still possible when asphaltenes interact with catalysts, especially acidic support catalysts, through the functional groups, e.g., the basic nitrogen species, just as they interact with adsorbents. And there is a possibility

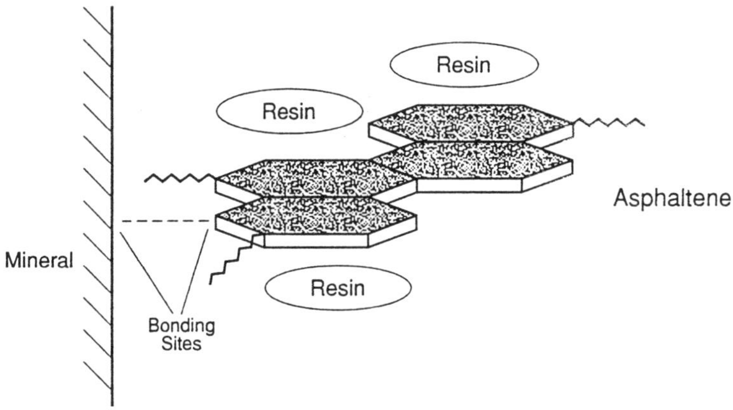

Figure 8-7. Schematic representation of the single-point adsorption of asphaltenes.

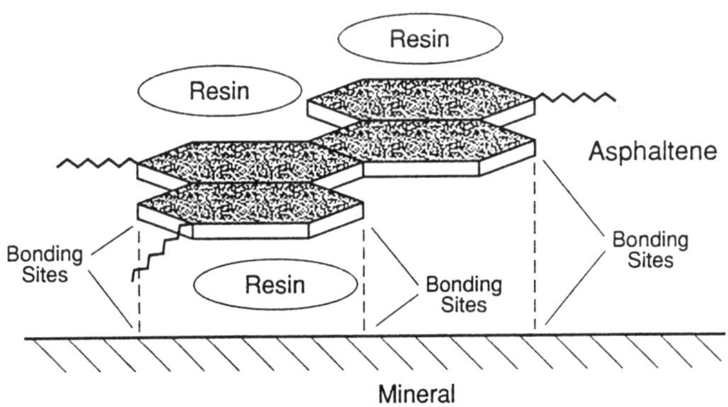

Figure 8-8. Schematic representation of the multipoint adsorption of asphaltenes.

for interaction of the asphaltene with the catalyst through the agency of a single functional group in which the remainder of the asphaltene molecule remains in the liquid phase (Figure 8-7) and has the potential to be converted to liquids. There is also a less desirable option in which the asphaltene reacts with the catalyst at several points of contact, causing immediate incompatibility on the catalyst surface (Figure 8-8) and thereby having the potential to block more of the catalyst surface as coke is laid down.

4.4 Deasphalting

The deasphalting process is perhaps the best example of a processes in which incompatibility of the asphaltenes is preferred. In fact, the deasphalting process can be classed as a carbon rejection (albeit nonthermal) process.

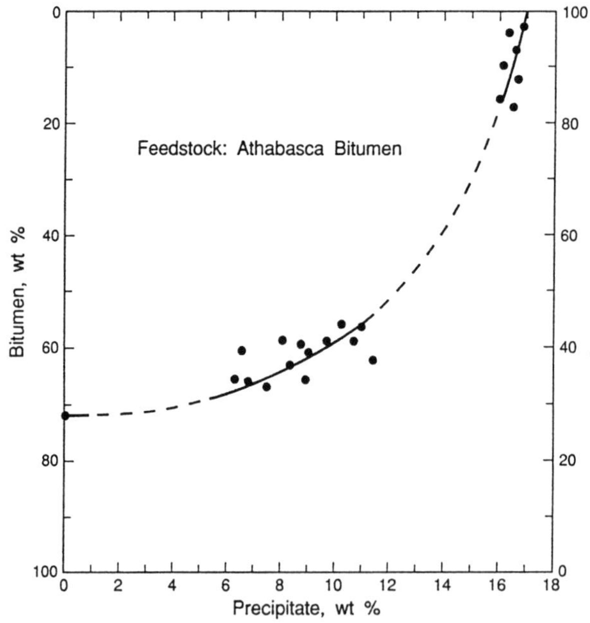

Figure 8-9. Variation of asphaltenes precipitate with the amount of hydrocarbon liquid added.

In the deasphalting process in which asphaltenes, and often resins, are discharged from the feedstock on addition of hydrocarbon liquids (Mitchell and Speight, 1973; Speight, 1991). The interesting point is that the amount of precipitate can be equated to the solubility parameter of the liquid phase (Mitchell and Speight, 1973) and offers the attractive option of predictability. In addition, the amount of precipitate varies with the amount of added paraffin liquid (Figure 8-9), which may be more representative of the situation that occurs in many reactors during refinery operations.

In a similar manner, the solubility parameter of the various fractions of petroleum can be estimated from the atomic hydrogen-to-carbon atomic ratio (Figure 8-10) (Speight, 1994). It is therefore possible to represent petroleum as a three-phase system (Figure 8-11) in which chemical and/or physical changes can be represented. Knowing the character of the petroleum, it becomes possible to map the parameters of the stable and unstable phases, thereby allowing some prediction of the onset of instability or incompatibility.

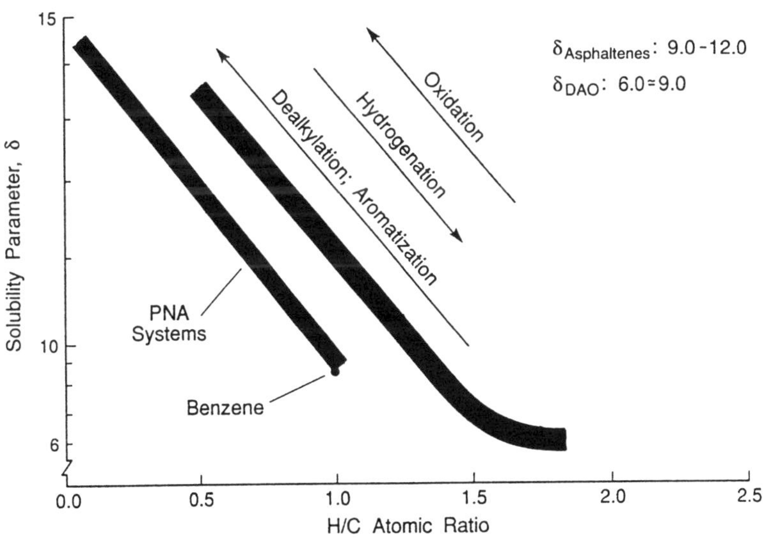

Figure 8-10. Estimation of solubility parameter from the hydrogen-to-carbon atomic ratio.

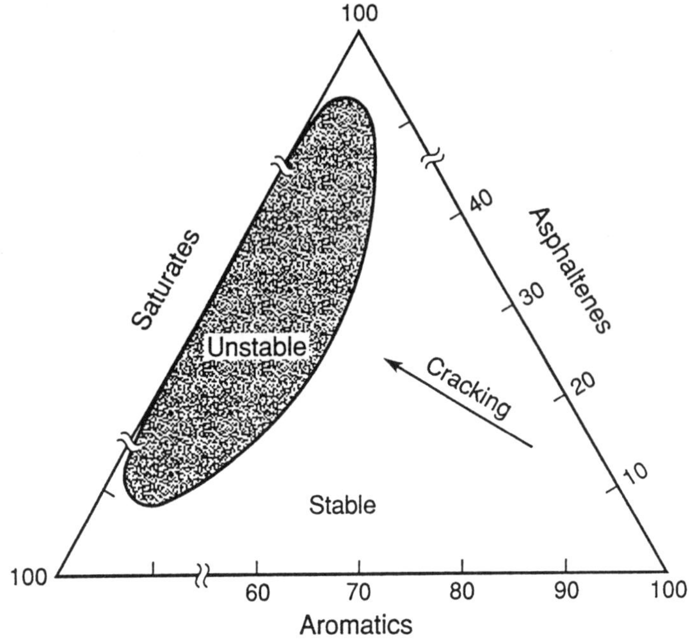

Figure 8-11. Representation of petroleum as a three-phase system for predicting incompatibility.

5.0 REFERENCES

Allan, D. E., Martinez, C. H., Eng. C. C., and Barton, W. J. 1983. Chem. Eng. Progr. 79(1): 85.

Barth, E. J. 1962. Asphalt: Science and Technology. Gordon and Breach, New York.

Batts, B. D., and Fathoni, A. Z. 1991. Energy & Fuels. 5: 5.

Benson, S. W., and Shaw, R. 1970. Organic Peroxides. Volume I. Wiley-Interscience, New York. Chapter 2.

Bjorseth, A. 1983. Handbook of Polycyclic Aromatic Hydrocarbons. Marcel Dekker Inc., New York.

Broome, D. C., and Wadelinn, F. A. 1973. In Criteria for Quality of Petroleum Products. J. P. Allinson (Editor). Halsted Press, Toronto. Chapter 13.

CRC. 1988. Determination of the Hydroperoxide Potential of Jet Fuels. Report No. 559. Coordinating Research Council Inc., Atlanta, GA.

Dias, J. R. 1987. Handbook of Polycyclic Hydrocarbons. Part A. Benzenoid Hydrocarbons. Elsevier, New York.

Dias, J. R. 1988. Handbook of Polycyclic Hydrocarbons. Part B. Polycyclic Isomers and Analogs of Benzenoid Hydrocarbons. Elsevier, New York.

Frankenfeld, J. W., Taylor, W. F., and Brinkman, D. W. 1983. Ind. Eng. Chem. Prod. Res. Dev. 22: 608.

Gary, J. G., and Handwerk, G. E. 1984. Petroleum Refining: Technology and Economics. Marcel Dekker Inc., New York.

Gray, M. R. 1994. Upgrading Petroleum Residues and Heavy Oils. Marcel Dekker Inc., New York.

Gruse, W. A., and Stevens, D. R. 1960. Chemical Technology of Petroleum. McGraw-Hill, New York. Chapter XII.

Hardy, D. R., and Wechter, M. A. 1990. Fuel. 69: 720.

Hartough, H. D., and Meisel, S. L. 1954 Compounds with Condensed Thiophene Rings. Wiley Interscience, New York.

Henderson, J. H., and Weber, L. 1965. J. Can. Pet. Tech. 4: 206.

Hoffman, H. L. 1992. In Petroleum Processing Handbook. J. J. McKetta (Editor). Marcel Dekker Inc., New York. p. 2.

Hoiberg, A. J. (Editor). 1964. Bituminous Materials: Asphalts, Tar, and Pitches. Volumes I, II, and III. Interscience Publishers, New York.

Kharasch, N. (Editor). 1961. Organic Sulfur Compounds, Vol. I. Pergamon Press, New York.

Kite, W. H., and Stephens, G. G. 1973. In Criteria for Quality of Petroleum Products. J. P. Allinson (Editor). Halsted Press, Toronto. Chapter 9.

Long, R. B. 1979. Preprints Div. Petrol. Chem. Am. Chem. Soc. 24(4): 891.

Long, R. B. 1981. In The Chemistry of Asphaltenes. J.W. Bunger and N. Li (Editors). Advances in Chemistry Series No. 195. American Chemical Society, Washington, DC. Chapter 2.

Magaril, R. Z., and Akensova, E. I. 1968. Int. Chem. Eng. 8: 727.

Magaril, R. Z., and Ramazaeva, L. F. 1969. Izv. Vyssh. Ucheb. Zaved. Neft Gaz. 12(1): 61.

Magaril, R. Z., Ramazaeva, L. F., and Aksenova, E. I. 1970. Khim. Tekhnol. Topliv. i Masel. 15(3): 15.

McKetta, J. J. (Editor). 1992. Petroleum Processing Handbook. Marcel Dekker Inc., New York.

Migita, T., Kosugi, M., Takayama, K., and Nakagawa, Y. 1973. Tetrahedron. 29: 51.

Mitchell, D. L., and Speight, J. G. 1973. Fuel. 52: 149.

Mochida, J., Sakanishi, K., and Fujitsu, H. 1986. Oil Gas J. November.

Mushrush, G. W. 1992. Fuel Sci. Technol. Int. 10: 1523.

Mushrush, G. W., and Speight, J. G. 1995. Petroleum Products: Stability and Incompatibility. Taylor & Francis, Washington, DC.

Mushrush, G. W., Pellenbarg, R. E., Hazlett, R. N., Morris, R. E., and Hardy, D. R. 1991. Fuel Sci. Technol. Int. 9: 1137.

Mushrush, G. W., Honeychuck, R. V., Stalick, W. M., Slone, J. E., Beal, E. J., and Hardy, D. R. 1995. Fuel Sci. Technol. Int. 13: 793.

Pedley, J. F., Hiley, R. W., and Hancock, R. A. 1987. Fuel. 66: 1645.

Pedley, J. F., Hiley, R. W., and Hancock, R. A. 1988. Fuel. 67: 1124.

Pope, J. G. C. 1973. In Modern Petroleum Technology. G. D. Hobson and W. Pohl (Editors). Applied Science Publishers, Barking Essex, England. Chapter 18.

Por, N. 1992. Stability Properties of Petroleum Products. Israel Institute of Petroleum and Energy, Tel Aviv, Israel.

Por, N., Brauch, R., Brodsky, N., and Diamant, R. 1988. Proceedings, 3rd International Conference on the Stability and Handling of Liquid Fuels. London.

Power, A. J., and Mathys, G. I. 1992. Fuel. 71: 903.

Rao, B. M. L., and Serrano, J. E. 1986. Fuel Sci. Technol. Int. 4: 483.

Rhoe, A., and de Blignieres, C. 1979. Hydrocarbon Process. 58(1): 131.

Ruzicka, D. J., and Nordenson, S. 1990. Fuel. 69: 710.

Schabron, J. F., and Speight, J. G. 1997. Preprints Div. Petrol. Chem. Am. Chem. Soc. 42: 386.

Schucker, R. C., and Keweshan, C. F. 1980. Preprints Div. Fuel Chem. Am. Chem. Soc. 25: 155.

Speight, J. G. 1991. The Chemistry and Technology of Petroleum. 2nd Edition. Marcel Dekker Inc., New York.

Speight, J. G. 1994. In Asphalts and Asphaltenes, 1. T. F. Yen and G. V. Chilingarian (Editors). Elsevier, Amsterdam, The Netherlands. Chapter 2.

Speight, J. G., and Moschopedis, S. E. 1979. Fuel Process. Technol. 2: 295.

Stavinoha, L. L., and Henry, C. P. (Editors). 1981. Distillate Fuel Stability and Cleanliness. Special Technical Publication No. 751. American Society for Testing and Materials, Philadelphia.

Sundberg, R. J. 1970. The Chemistry of Indoles, Academic Press Inc., New York, p. 9.

Traxler, R. N. 1961. Asphalt: Its Composition, Properties, and Uses. Reinhold Publishing Corp., New York.

Wallace, T. J. 1964. Advances in Petroleum Chemistry and Refining, Wiley-Interscience, New York. p. 353.

Walmsley, A. G. 1973. In Modern Petroleum Technology. G. D. Hobson and W. Pohl (Editors). Applied Science Publishers, Barking Essex, England. Chapter 17.

Watkins, J. M., Mushrush, G. W., Hazlett, R. N., and Beal, E. J. 1989. Energy and Fuels. 3: 231.

Wiehe, I. A. 1993. Ind. Eng. Chem. Res. 32: 2447.

Glossary

This Glossary of terms is presented here to aid the reader of this book who may be unfamiliar with some of the terminology used by petroleum scientists, engineers, and refiners and no claims are made in regard to its completeness. The terms presented are those commonly in use in the petroleum industry.

ABN separation: a method of fractionation by which petroleum is separated into acidic, basic, and neutral constituents.

Acid catalyst: a catalyst having acidic character; the aluminas are examples of such catalysts.

Acid deposition: acid rain; a form of pollution depletion in which pollutants, such as nitrogen oxides and sulfur oxides, are transferred from the atmosphere to soil or water; often referred to as atmospheric self-cleaning. The pollutants usually arise from the use of fossil fuels.

Acid gas: hydrogen sulfide (H_2S) or carbon dioxide ($C0_2$).

Acid rain: precipitation of anthropogenic acids and other acidic chemicals from the atmosphere to the land and water (see **Acid deposition**).

Acidity: the capacity of an acid to neutralize a base such as a hydroxyl ion (OH^-).

Add-on control methods: the use of devices that remove refinery process emissions after they are generated but before they are discharged to the atmosphere.

Adsorption: transfer of a substance from a solution to the surface of a solid.

Air pollution: toxic gases and particulate matter introduced into the atmosphere, principally as a result of human activity.

Alkalinity: the capacity of a base to neutralize the hydrogen ion (H^+).

Alkylate: the product of an alkyation (q.v.) process.

Alkylation: in the petroleum industry, a process by which olefins are combined with branched-chain hydrocarbons such as *iso*-butane in the presence of a catalyst.

Alumina: Al_2O_3; used in separation methods as an adsorbent and in refining as a catalyst.

American Society for Testing and Materials (ASTM): the official organization in the United States for designing standard tests for petroleum and other industrial products.

Amine washing: a method of gas cleaning whereby acidic impurities such as hydrogen sulfide and carbon dioxide are removed from the gas stream by washing with an amine (usually an alkanolamine).

Aniline point: the temperature, usually expressed in °F, above which equal volumes of a petroleum product are completely miscible; a qualitative indication of the relative proportions of paraffins in a petroleum product which are miscible with aniline only at higher temperatures; a high aniline point indicates low aromatics.

Anthracene oil: the heaviest distillable coal tar fraction, with distillation range 270-400°C (520-750°F), containing creosote oil, anthracene, phenanthrene, carbazole, and so on.

API gravity: a measure of the *lightness* or *heaviness* of petroleum that is related to density and specific gravity.

Aromatics: benzene and substituted benzene, naphthalene and substituted naphthalene, phenanthrene and substituted phenanthrene, as well as the higher condensed ring systems.

Ash: the noncombustible residue remaining after complete combustion of a petroleum sample; the final form of the mineral matter present in petroleum.

Ash analysis: percentages of inorganic oxides present in an ash sample; indirectly equatable to the mineral matter and inorganic constituents of petroleum.

Asphalt: a nonvolatile product of petroleum refining used for road construction.

Asphaltene association factor: the number of individual asphaltene species that associate in nonpolar solvents as measured by molecular weight methods; the molecular weight of asphaltenes in toluene divided by the molecular weight in a polar nonassociating solvent, such as dichlorobenzene, pyridine, or nitrobenzene.

Asphaltenes: the fraction of petroleum that is insoluble in a liquid hydrocarbon such as *n*-pentane or *n*-heptane.

Associated molecular weight: the molecular weight of asphaltenes in an associating (nonpolar) solvent, such as toluene.

Atmospheric distillation: distillation (q.v.) at atmospheric pressure.

Atmospheric equivalent boiling point (AEBP): a mathematical estimate the boiling point at atmospheric pressure of nonvolatile fractions of petroleum.

Atmospheric residuum: the portion of petroleum that boils above 350°C (660°F).

Baghouse: a filter system for the removal of particulate matter from gas streams; so called because of the similarity of the filters to coal bags.

Barrel: liquid volume measure equal to 42 US gals, commonly used in measuring petroleum or petroleum products.

Biomass: biological organic matter.

Bitumen: a naturally occurring material that is petroleum-like in nature but that is immobile at room temperature; similar in properties to asphalt (q.v.).

Bituminous sand: see **Tar Sand**.

Bond dissociation energy: the energy required to rupture a chemical bond; a measure of the amount of heat (energy) required to refine certain feedstocks.

Bottom ash: ash that occurs at the bottom of a (for example, coal) combustor.

Bronsted acid: a chemical species that can act as a source of protons.

Bronsted base: a chemical species that can accept protons.

Bunker C oil: see **No. 6 fuel oil**.

Carbenes: the fraction of petroleum that is insoluble in benzene (or toluene) but soluble in pyridine (or carbon disulfide); also rifles used to shoot bison.

Carboids: the fraction of petroleum which is insoluble in benzene (or toluene) and in soluble in pyridine (or carbon disulfide).

Carbonate washing: processing using a mild alkali (e.g., potassium carbonate) process for emission control by the removal of acid gases from gas streams.

Carbon dioxide: gas (CO_2) formed by the complete combustion of petroleum (hydrocarbon) products.

Carbon-forming propensity: see **Carbon residue**.

Carbon monoxide: gas (CO) formed by incomplete combustion of petroleum (hydrocarbon) products.

Carbon rejection: an upgrading process in which coke is produced, e.g., coking.

Carbon residue: the amount of carbonaceous residue remaining after thermal decomposition of petroleum, a petroleum fraction, or a petroleum product in a limited amount of air; also called the *coke-* or *carbon-forming propensity*; often prefixed by the term *Conradson* or *Ramsbottom* in reference to the inventor of the respective test.

Catalyst: a chemical agent that, when added to a reaction (process), enhances the conversion of a feedstock without being consumed in the process.

Catalytic cracking : a process in which heavy gas oils are converted into products of higher volatility by contact of higher molecular weight hydrocarbon with the hot catalyst.

Cetane number: a number indicating the ignition quality of diesel fuel; a high cetane number represents a short ignition delay time; the ignition quality of diesel fuel can also be estimated from the following formula:

diesel index = (aniline point in °F x API gravity)100

Chelating agents: complex-forming agents having the ability to solubilize heavy metals.

Chemical waste: any solid, liquid, or gaseous material discharged from a process that may pose substantial hazards to human health and the environment.

Clay: silicate minerals that also usually contain aluminum and have particle sizes of less than 0.002 micron; used in separation methods as an adsorbent and in refining as a catalyst.

Coal: an organic rock.

Coke: solid product produced from petroleum during thermal processes; also produced by the carbonization of coal

Coking: a process for the thermal conversion of petroleum in which gaseous, liquid, and solid (coke) products are formed.

Combustible liquid: a liquid with a flash point in excess of 37.8°C (100°F) but below 93.3°C (200°F).

Composition: the general chemical makeup of petroleum.

Composition map: a means of illustrating the chemical makeup of petroleum using chemical and/or physical property data.

Conradson carbon residue: see **Carbon residue**.

Contaminant: a substance that causes deviation from the normal composition of an environment.

Conversion: the thermal treatment of petroleum that results in the formation of new products by altering of the original constituents.

Cracked residua: residua that have been subjected to temperatures above 350°C (660°F) during the distillation process.

Cracking: the thermal processes by which the constituents of petroleum are converted to lower molecular weight products.

Cracking temperature: the temperature (350°C; 660°F) at which the rate of thermal decomposition of petroleum constituents becomes significant.

Crude oil: an alternative name for petroleum.

Cyclone: a device for extracting dust from industrial waste gases that takes the form of an inverted cone into which the contaminated gas enters tangentially from the top; the gas is propelled down a helical pathway, and the dust particles are deposited by means of centrifugal force onto the wall of the scrubber.

Deasphaltened oil: the fraction of petroleum after the asphaltenes have been removed.

Deasphaltening: removal of a solid powdery asphaltene fraction from petroleum by the addition of a low-boiling liquid hydrocarbon such as *n*-pentane or *n*-heptane under ambient conditions.

Deasphalting: removal of the asphaltene fraction from petroleum by the addition of a low-boiling hydrocarbon liquid such as *n*-pentane or *n*-heptane; more correctly the removal of asphalt (tacky, semisolid) from petroleum (as occurs in a refinery asphalt plant) by the addition of liquid propane or liquid butane under pressure.

Dehydrating agents: substances capable of removing water (drying, q.v.) or the elements of water from another substance.

Delayed coking: a coking process in which the thermal reactions are allowed to proceed to completion to produce gaseous, liquid, and solid (coke) products.

Density: the mass (or weight) of a unit volume of any substance at a specified temperature; see also **Specific gravity**.

Devolatilized fuel: smokeless fuel; coke that has been reheated to remove all of the volatile material.

Diagenesis: the concurrent and consecutive chemical reactions that result in the formation of petroleum from marine sediments.

Diesel cycle: a repeated succession of operations representing the idealized working behavior of the fluids in a diesel engine.

Diesel index: see **Cetane number**.

$$\text{diesel index} = (\text{aniline point in } ^{\circ}\text{F x API gravity})100$$

Distillation: a process for separating liquids with different boiling points.

Domestic heating oil: see **No. 2 fuel oil**.

Donor solvent process: a conversion process in which hydrogen donor solvent is used in place of or to augment hydrogen.

Drying: removal of a solvent or water from a chemical substance; also the removal of solvent from a liquid or suspension.

Ebullated bed: a catalyst bed in a suspended state in the reactor by means of a feedstock recirculation pump, which pumps the feedstock upward at sufficient speed to expand the catalyst bed at approximately 35% above the settled level.

Effluent: any contaminating substance, usually a liquid, that enters the environment via a domestic industrial, agricultural, or sewage plant outlet.

Electrostatic precipitators: devices used to trap fine dust particles (usually in the size range 30–60 microns) that operate on the principle of imparting an electric charge to particles in an incoming airstream and then collecting the particles on an oppositely charged plate across a high-voltage field.

Emission control: use of gas cleaning processes to reduce emissions.

Emission standard: the maximum amount of a specific pollutant permitted to be discharged from a particular source in a given environment.

Emulsion breaking: the settling or aggregation of colloidal-sized emulsions from suspension in a liquid medium.

End-of-pipe emission control: use of specific emission control processes to clean gases after production of the gases.

Endothermic reaction: a process in which heat is absorbed.

Energy: the capacity of a body or system to do work, measured in joules (SI units); also the output of fuel sources.

Energy from biomass: the production of energy from biomass (q.v.).

Entrained bed: a bed of solid particles suspended in a fluid (liquid or gas) at such a rate that some of the solid is carried over (entrained) by the fluid.

Ethanol: see Ethyl alcohol.

Ethyl alcohol (ethanol or grain alcohol): an inflammable organic compound (C_2H_5OH) formed during fermentation of sugars; used as an intoxicant and as a fuel.

Evaporation: a process for concentrating nonvolatile solids in a solution by boiling off the liquid portion of the waste stream.

Exothermic reaction: a process in which heat is evolved.

Expanding clays: clays that expand or swell on contact with water, e.g., montmorillonite.

Fabric filters: filters made from fabric materials and used for removing particulate matter from gas streams (see Baghouse).

FCC: fluid catalytic cracking.

Feedstock: petroleum as it is fed to the refinery; a refinery product that is used as the raw material for another process; the term is also generally applied to raw materials used in other industrial processes.

Filtration: the use of an impassable barrier to collect solids but allows liquids to pass.

Fixed bed: stationary bed (of catalyst) used to accomplish a process; see **Fluid bed**.

Flammability range: the range of temperature over which a chemical is flammable.

Flammable: the property of a substance of burning readily.

Flammable liquid: a liquid having a flash point below 37.8°C (100°F).

Flammable solid: a solid that can be ignited by friction or heat remaining from its manufacture, or that may cause a serious hazard if ignited.

Flash point: the lowest temperature to which the product must be heated under specified conditions to give off sufficient vapor to form a mixture with air that can be ignited momentarily by a flame.

Flue gas desulfurization (FGD or scrubbing): the removal of sulfur oxides and nitrogen oxides from stack gas.

Fluid bed: an agitated bed of inert granular material used to accomplish a process in which the agitation of the bed resembles the motion of a fluid.

Flexicoking: a modification of the fluid coking process in which a gasifier adjoins the burner/regenerator is used to convert excess coke to a clean fuel gas.

Fluid catalytic cracking: cracking in the presence of a fluidized bed of catalyst.

Fluid coking: a continuous fluidized solids process in which feed is cracked thermally over heated coke particles in a reactor vessel to gas, liquid products, and coke.

Fly ash: particulate matter produced from mineral matter in coal that is converted during combustion to finely divided inorganic material and that emerges from the combustor in the gases.

Fossil fuel resources: a gaseous, liquid, or solid fuel material formed in the ground by chemical and physical changes (Diagenesis, q.v.) in plant and animal residues over geological time; natural gas, petroleum, coal, and oil shale.

Fractional composition: the composition of petroleum as determined by fractionation (separation) methods.

Fractionation: the separation of petroleum into the constituent fractions using solvent or adsorbent methods; chemical agents such as sulfuric acid may also be used.

Fuel oil: also called **heating oil** is a distillate product that covers a wide range of properties; see also **No. 1 to No. 4 fuel oils**.

Gas cleaning: gas treatment to remove contaminants such as entrained solids, carbon dioxide (q.v.), carbon monoxide (q.v.), hydrogen sulfide, sulfur oxides, and nitrogen oxides from refinery gas.

Gaseous pollutants: gases released into the atmosphere that act as primary or secondary pollutants.

Gasohol: a motor vehicle fuel comprising between 80–90% unleaded gasoline and 10–20% ethanol; see also **Ethyl alcohol**.

Grain alcohol: see **Ethyl alcohol**.

Granular bed filters: equipment that uses a bed of a separate, closely packed solids as the separation medium.

Greenhouse effect: warming of the earth due to entrapment of the sun's energy by the atmosphere.

Greenhouse gases: gases that contribute to the greenhouse effect (q.v.).

Gum: an insoluble, tacky semisolid material formed as a result of storage instability and/or the thermal instability of petroleum and petroleum products.

Heating oil: see **Fuel oil**.

Heavy fuel oil: fuel oil having a high density and viscosity; generally residual fuel oil such as No. 5 and No 6. fuel oil (q.v.)

Heavy petroleum: petroleum having an API gravity less than 20°.

Heteroatom compounds: chemical compounds that contain nitrogen and/or oxygen and/or sulfur and/or metals bound within their molecular structure(s).

Hot filtration test: a test for the stability of a petroleum product.

Hydrocarbon compounds: chemical compounds containing only carbon and hydrogen.

Hydrocarbon resources: resources such as petroleum and natural gas that can produce naturally-occurring hydrocarbons without the application of conversion processes.

Hydrocarbon-producing resource: a resource such as coal and oil shale (kerogen) that produces *derived* hydrocarbons on the application of conversion processes; the hydrocarbons so produced are not naturally occurring materials.

Hydroconversion: a term often applied to hydrocracking (q.v.).

Hydrocracking: a catalytic high-pressure, high-temperature process for the conversion of petroleum feedstocks in the presence of fresh and recycled hydrogen; carbon-carbon bonds are cleaved in addition to the removal of heteroatomic species.

Hydrocracking catalyst: a catalyst used for hydrocracking, which typically contains separate hydrogenation and cracking functions.

Hydrodenitrogenation: the removal of nitrogen by hydrotreating (q.v.).

Hydrodesulfurization: the removal of sulfur by hydrotreating (q.v.).

Hydrogen addition: an upgrading process in the presence of hydrogen, e.g. hydrocracking.

Hydrogenation: a chemical reaction in which hydrogen is added to a substance.

Hydrophilic: possessing polar surfaces that are readily wetted by water; *water-loving*.

Hydrophobic: possessing nonpolar surfaces that are not wetted by water; *water-hating*.

Hydroprocessing: a term often equally applied to hydrotreating (q.v.) and to hydrocracking (q.v.); also often collectively applied to both.

Hydropyrolysis: a short residence time, high-temperature process using hydrogen.

Hydrotreating: the removal of heteroatomic (nitrogen, oxygen, and sulfur) species by treatment of a feedstock or product at relatively low temperatures in the presence of hydrogen.

Hydrovisbreaking: a noncatalytic process, conducted under conditions similar to those of visbreaking, that involves treatment with hydrogen to reduce the viscosity of the feedstock and produce more stable products than are possible with visbreaking.

Ignitability: characteristic of liquids whose vapors are likely to ignite in the presence of an ignition source; also characteristic of non liquids that may catch fire as a result of friction or contact with water and that burn vigorously.

Incompatibility: the *immiscibility* of petroleum products and also of different crude oils, which is often reflected in the formation of a separate phase after mixing and/or storage.

Instability: inability of a petroleum product to exist for periods of time without being changed.

Ion exchange: a means of removing cations or anions from solution onto a solid resin.

Isomerization: the conversion of a *normal* (straight-chain) paraffin hydrocarbon into an *iso* (branched-chain) paraffin hydrocarbon having the same atomic composition.

Kaolinite: a clay mineral formed by hydrothermal activity at the time of rock formation or by chemical weathering of rocks with a high feldspar content; usually associated with intrusive granite rocks with a high feldspar content.

Kerogen: a complex macromolecule often regarded as a precursor of petroleum but generally formed later in the diagenetic sequence after *protopetroleum*.

Kerosene (kerosine): a fraction of petroleum that was initially sought as an illuminant in lamps; a precursor of diesel fuel.

Lewis acid: a chemical species that can accept an electron pair from a base.

Lewis base: a chemical species that can donate an electron pair.

Light petroleum: petroleum having an API gravity greater than 20°.

Liquefied petroleum gas (LPG): a mixture of liquid propane and liquid butane.
LPG: see **Liquefied petroleum gas**.

Maltenes: deasphaltened oil.
Marine sediment: the organic biomass from which petroleum is derived
Marsh: an area of spongy, waterlogged ground with large numbers of surface water pools; marshes usually result from (1) an impermeable underlying bedrock; (2) surface deposits of glacial boulder clay; (3) a basin-like topography from which natural drainage is poor; (4) very heavy rainfall in conjunction with a low evaporation rate; (5) low-lying land, particularly at estuarine sites at or below sea level.
Methanol: see **Methyl alcohol**.
Methyl alcohol (methanol; wood alcohol): a colorless, volatile, inflammable, and poisonous alcohol (CH_3OH) traditionally formed by destructive distillation of wood or, more recently, synthetic distillation in chemical plants.
Mica: a complex aluminum silicate mineral that is transparent, tough, flexible, and elastic.
Micelle: the structural entity by which asphaltenes are dispersed in petroleum.
Microcarbon residue: the carbon residue determined using a themogravimetric method. See also **Carbon residue**.
Mineral oil: the older term for petroleum; the term was introduced in the nineteenth century as a means of differentiating petroleum (rock oil) from whale oil, which, at the time, was the predominant illuminant for oil lamps.
Minerals: naturally occurring inorganic solids with well-defined crystalline structures.
Mitigation: identification, evaluation, and elimination of potential impacts of a process product or by-product.
Modified naphtha insolubles (MNI): an insoluble fraction obtained by adding naphtha to petroleum; usually the naphtha is modified by adding paraffinic constituents; the fraction might be equated to asphaltenes *if* the naphtha is equivalent to *n*-heptane, but usually it is not.

Naft: pre-Christian era (Greek) term for naphtha (q.v.).
Naphtha: the volatile fraction of petroleum that is used as a solvent or as a precursor to gasoline.
Naphthenes: cycloparaffins.
Natural gas: gaseous components that often occur in reservoirs with petroleum; when pure, natural gas has a heat content in excess of 1000 Btu/ft^3 and

consists mainly of methane but also contains smaller amounts of the C_2-C_4 hydrocarbons as well as nitrogen, carbon dioxide, and hydrogen sulfide.

Neutralization: a process for reducing the acidity or alkalinity of a waste stream by mixing acids and bases to produce a neutral solution; also known as pH adjustment.

No. 1 fuel oil: very similar to kerosene (q.v.); used in burners where vaporization before burning is usually required and a clean flame is specified.

No. 2 fuel oil: also called **domestic heating oil**; has properties similar to those of diesel fuel and heavy jet fuel; used in burners where complete vaporization is not required before burning.

No. 4 fuel oil: a light industrial heating oil used where preheating is not required for handling or burning; there are two grades of No. 4 fuel oil, differing in safety (flash point) and flow (viscosity) properties.

No. 5 fuel oil: a heavy industrial fuel oil that requires preheating before burning.

No. 6 fuel oil: a heavy fuel oil more commonly known as **Bunker C oil** when it is used to fuel ocean-going vessels; preheating is always required for burning this oil.

Octane number: a number indicating the anti-knock characteristics of gasoline.

Oil sand: see **Tar Sand**.

Oil shale: a fine-grained impervious sedimentary rock which contains an organic material called kerogen.

Organic sedimentary rocks: rocks containing organic material such as residues of plant and animal remains (decay).

Oxidation: a process that can be used for the treatment of a variety of inorganic and organic substances.

Particulate matter: particles in the atmosphere or in a gas stream that may be organic or inorganic and originate from a wide variety of sources and processes.

Permeability: the ease of flow of the water through rock.

Petrolenes: the fraction of the maltenes that boils below 345°C (650°F).

Petroleum (crude oil): a naturally occurring mixture of gaseous, liquid, and solid hydrocarbon compounds usually found trapped deep underground beneath impermeable cap rock and above a lower dome of sedimentary rock such as shale; most petroleum reservoirs occur in sedimentary rocks of marine, deltaic, or estuarine origin.

Petroleum refining: a complex sequence of events that result in the production of a variety of products.

Petroporphyrins: see **Porphyrins**.

Phase separation: the formation of a separate phase that is usually the prelude to coke formation during a thermal process; the formation of a separate phase as a result of the instability or incompatibility of petroleum and petroleum products.

pH adjustment: neutralization.

Pipeline gas: a methane-rich gas with a heat content in excess of 950 Btu/ft^3 compressed to 1000 psi.

Pipestill gas: the most volatile fraction that contains most of the gases that are generally dissolved in the crude; also known as pipestill light ends.

Pitch: the nonvolatile product of the thermal decomposition of coal.

PNA: polynuclear aromatic compound (q.v.).

Polar aromatics: resins; the constituents of petroleum that are predominantly aromatic in character and contain polar (nitrogen, oxygen, and sulfur) functions in their molecular structure(s).

Pollution: the introduction into the land, water, and air systems of a chemical or chemicals that are not indigenous to these systems or the introduction into the land, water, and air systems of indigenous chemicals in greater than natural amounts.

Polymerization: the combination of two olefin molecules to form a higher molecular weight paraffin.

Polynuclear aromatic compound: an aromatic compound having two or more fused benzene rings, e.g., naphthalene, phenanthrene.

Porosity: the percentage of rock volume available to contain water or other fluid.

Porphyrins: organometallic constituents of petroleum that contain vanadium or nickel; the degradation products of chorophylls that were included in the protopetroleum.

Possible reserves: reserves for which there is a greater degree of uncertainty but about which there is some information.

Potential reserves: reserves for which there is geological information about the types of sediments where resources are likely to occur; they are considered to represent an educated guess.

Primary structure: the chemical sequence of atoms in a molecule.

Probable reserves: mineral reserves that are nearly certain but about which a slight doubt exists.

Protopetroleum: a generic term used to indicate the initial product formed when changes have occurred in the precursors of petroleum.

Proved reserves: mineral reserves that have been positively identified as recoverable with current technology.

Pyrolysis: exposure of a feedstock to high temperatures in an oxygen-poor environment.

Pyrophoric: catching fire spontaneously in air without an ignition source.

Ramsbottom carbon residue: see **Carbon residue**.

Raw materials: minerals extracted from the earth prior to any refining or treating.

Recycling: the use or reuse of chemical waste as an effective substitute for a commercial products or as an ingredient or feedstock in an industrial process.

Refinery: a complex network of integrated unit processes for producing a variety of products from petroleum.

Refinery gas: a gas (or a gaseous mixture) produced as a result of refining operations.

Refining: the process(es) by which petroleum is distilled and/or converted by application of physical and chemical processes to form a variety of products.

Reformate: the liquid product of a reforming process.

Reforming: the conversion of hydrocarbons with low octane numbers (q.v.) into hydrocarbons having higher octane numbers; e.g., the conversion of a *n*-paraffin into a iso-paraffin.

Reformulated gasoline (RFG): gasoline designed to mitigate smog production and to improve air quality by limiting the emission levels of certain chemical compounds such as benzene and other aromatic derivatives.

Reid vapor pressure: a measure of the volatility of liquid fuels, especially gasoline.

Renewable energy sources: solar, wind, and other nonfossil fuel energy sources.

Reserves: well-identified resources that can be profitably extracted and utilized with existing technology.

Reservoir: a domain where a fossil fuel or a pollutant may reside for an indeterminate time.

Residual fuel oil: fuel oil obtained by blending residual products from various refining processes with suitable diluents (usually middle distillates) to obtain the required fuel oil grades.

Residual oil: see **Residuum**.

Residuum or resid (pl. residua, resids): the nondistillable fraction of petroleum.

Resins: the fraction of deasphaltened oil that is insoluble in liquid propane but soluble in *n*-heptane.

Resource: the total amount of a commodity (usually a mineral but can include nonminerals such as water and petroleum) that has been estimated to be ultimately available.

Run-of-the-river reservoirs: reservoirs with a large rate of flow-through compared to their volume.

Sand: a course granular mineral mainly comprising quartz grains that is derived from the chemical and physical weathering of rocks rich in quartz, notably sandstone and granite.

Sandstone: a sedimentary rock formed by compaction and cementation of sand grains; can be classified according to the mineral composition of the sand and cement.

SARA separation: a method of fractionation by which petroleum is separated into saturate, aromatic, resin, and asphaltene fractions.

Saturates: paraffins and cycloparaffins (naphthenes).

Scrubber: see **Flue gas desulfurization**.

Secondary structure: the ordering of the atoms of a molecule in space relative to each other.

Sediment: an insoluble solid formed as a result of storage instability and/or thermal instability of petroleum and petroleum products.

Sedimentary strata: strata that typically consist of mixtures of clay, silt, sand, organic matter, and various minerals.

Separation process: an upgrading process in which the constituents of petroleum are separated, usually without thermal decomposition, e.g., distillation and deasphalting.

Slime: a name used for petroleum in ancient texts.

Sludge: a semisolid to solid product that results from the storage instability and/or thermal instability of petroleum and petroleum products.

Slurry hydroconversion process: a process in which the feedstock comes in contact with hydrogen under pressure in the presence of a catalytic coke-inhibiting additive.

Slurry phase reactors: tanks into which wastes, nutrients, and microorganisms are placed.

Solubility parameter: a measure of the solvent power and polarity of a solvent.

Solvent extraction: a process for separating liquids by mixing the stream with a solvent that is immiscible with part of the waste but that will extract certain components of the waste stream.

Spontaneous ignition: ignition of a fuel, such as coal, under normal atmospheric conditions; usually induced by climatic conditions.

Specific gravity: the mass (or weight) of a unit volume of any substance at a specified temperature compared to the mass of an equal volume of pure water at a standard temperature; see also **Density**.

Stabilization: the removal of volatile constituents from a higher boiling fraction or product (stripping); the production of a product that, to all intents and purposes, does not undergo any further reaction when exposed to the air.

Steam cracking: a conversion process in which the feedstock is treated with superheated steam.

Storage stability (or storage instability): the ability (inability) of a liquid to remain in storage over extended periods of time without appreciable deterioration as measured by gum formation and deposition of insoluble material (sediment).

Straight-run products: products obtained from a distillation unit and used without further treatment.

Strata: layers including the solid iron-rich inner core, molten outer core, mantle, and crust of the earth.

Stripping: a means of separating volatile components from less volatile ones in a liquid mixture by the partitioning of the more volatile materials to a gas phase of air or steam (see **Stabilization**).

Synthetic crude oil: a mixture of hydrocarbons obtained by coal liquefaction, by conversion of oil shale kerogen, or by conversion of tar sand bitumen.

Tar: a volatile liquid or semisolid product of the thermal decomposition of coal (see also **Pitch**); also, a name used for petroleum in ancient texts.

Tar sand: a bituminous sand or sandstone impregnated with a viscous bituminous material; also called oil sand or bituminous sand.

Tertiary structure: the three-dimensional structure of a molecule.

Thermal coke: the carbonaceous residue formed as a result of a noncatalytic thermal process; the Conradson carbon residue; the Ramsbottom carbon residue.

Thermal stability (thermal instability): the ability (inability) of a liquid to withstand relatively high temperatures for short periods of time without the formation of carbonaceous deposits (sediment or coke).

Trace elements: elements that occur at very low levels in a given system, such as nickel and vanadium.

Treatment: any method, technique, or process that changes the physical and/or chemical character of petroleum.

Ultimate analysis: elemental composition; the analytical percentage by weight of carbon, hydrogen, nitrogen, oxygen, and sulfur in petroleum or a petroleum fraction or product.

Unassociated molecular weight: the molecular weight of asphaltenes in an nonassociating (polar) solvent, such as dichlorobenzene, pyridine, or nitrobenzene.

Unstable: usually refers to a petroleum product that has more volatile constituents present or refers to the presence of olefin and other unsaturated constituents.

Upgrading: the conversion of petroleum to value-added salable products.

Vacuum distillation: distillation (q.v.) under reduced pressure.

Vacuum residuum: the portion of petroleum that boils above a selected temperature such as 510°C (950°F) or 565°C (1050°F).

Visbreaking: a process for reducing the viscosity of heavy feedstocks by controlled thermal decomposition.

Viscosity: a measure of the ability of a liquid to flow or a measure of its resistance to flow.

Wet scrubbers: devices in which a countercurrent spray liquid is used to remove impurities and particulate matter from a gas stream.

Wood alcohol: see **Methyl alcohol**.

Zeolite: a crystalline aluminosilicate used as a catalyst and having a particular chemical and physical structure.

Index

—A—

—S—

—T—